An Introduction to the Physics of Nuclei and Particles

RICHARD A. DUNLAP
Dalhousie University

BROOKS/COLE
CENGAGE Learning™

Australia • Brazil • Japan • Korea • Mexico • Singapore • Spain • United Kingdom • United States

BROOKS/COLE
CENGAGE Learning

An Introduction to the Physics of Nuclei and Particles
Richard A. Dunlap

Publisher: David Harris

Acquisitions Editor: Chris Hall

Developmental Editor: Rebecca Heider

Editorial Assistant: Seth Dobrin

Marketing Manager: Kelley McAllister

Marketing Assistant: Sandra Perin

Advertising Project Manager: Stacey Purviance

Production Project Manager: Belinda Krohmer

Print/Media Buyer: Doreen Suruki

Permissions Editor: Kiely Sexton

Technology Project Manager: Sam Subity

Production Service: Nesbitt Graphics, Inc.

Cover Designer: Bill Stanton

Cover Image: © Lawrence Berkeley/Photo Researchers, Inc.

Cover Printer: Webcom Limited

Compositor: International Typesetting and Composition

Printer: Webcom Limited

For product information and technology assistance, contact us at **Cengage Learning Customer & Sales Support, 1-800-354-9706**

For permission to use material from this text or product, submit all requests online at **cengage.com/permissions**
Further permissions questions can be emailed to **permissionrequest@cengage.com**

Library of Congress Control Number: 2002115597

ISBN-13: 978-0-534-39294-9

ISBN-10: 0-534-39294-6

Brooks/Cole
10 Davis Drive, Belmont
CA 94002-3098
USA

Cengage Learning is a leading provider of customized learning solutions with office locations around the globe, including Singapore, the United Kingdom, Australia, Mexico, Brazil, and Japan. Locate your local office at: **international.cengage.com/region**

Cengage Learning products are represented in Canada by Nelson Education, Ltd.

For your course and learning solutions, visit **academic.cengage.com**

Purchase any of our products at your local college store or at our preferred online store **www.ichapters.com**

Printed in the United States of America
2 3 4 5 6 7 11 10 09 08 07

for Ewa

Contents

Preface   xi

Part I   Introduction   1

CHAPTER 1
Basic Concepts   3

1.1   Introduction   3
1.2   Terminology and Definitions   4
1.3   Units and Dimensions   5
1.4   Sources of Information   6

CHAPTER 2
Particles and Interactions   7

2.1   Classification of Subatomic Particles   7
2.2   Classification and Ranges of Interactions   9
2.3   Conservation Laws   9

Part II   Nuclear Properties and Models   11

CHAPTER 3
Nuclear Composition and Size   13

3.1   Composition of the Nucleus   13
3.2   Rutherford Scattering   16

3.3   Charge Distribution of the Nucleus   19
3.4   Mass Distribution of the Nucleus   23

CHAPTER 4
Binding Energy and the Liquid Drop Model   26

4.1   Definition and Properties of the Nuclear
       Binding Energy   26
4.2   The Liquid Drop Model   28
4.3   Beta Stability   31
4.4   Nucleon Separation Energies   37

CHAPTER 5
The Shell Model   42

5.1   Overview of Atomic Structure   42
5.2   Evidence for Nuclear Shell Structure   45
5.3   The Infinite Square Well Potential   50
5.4   Other Forms of the Nuclear Potential   54
5.5   Spin-Orbit Coupling   56
5.6   Nuclear Energy Levels   58

CHAPTER 6
Properties of the Nucleus   62

6.1   Ground State Spin and Parity   62
6.2   Excited Nuclear States   65
6.3   Mirror Nuclei   67
6.4   Electromagnetic Moments of the Nucleus   67
6.5   Electric Quadrupole Moments   69
6.6   Magnetic Dipole Moments   71
6.7   An Overview of the Collective Model   75

Part III   Nuclear Decays and Reactions   81

CHAPTER 7
General Properties of Decay Processes   83

7.1   Decay Rates and Lifetimes   83
7.2   Quantum Mechanical Considerations   88
7.3   Radioactive Dating   90

CHAPTER 8
Alpha Decay   94

8.1   Energetics of Alpha Decay   94
8.2   Theory of Alpha Decay   97
8.3   Angular Momentum Considerations   102

CHAPTER 9
Beta Decay   106

9.1   Energetics of Beta Decay   106
9.2   Fermi Theory of Beta Decay   109
9.3   Fermi-Kurie Plots   114
9.4   Allowed and Forbidden Transitions   116
9.5   Parity Violation in Beta Decay   120
9.6   Double Beta Decay   122

CHAPTER 10
Gamma Decay   128

10.1   Energetics of Gamma Decay   128
10.2   Classical Theory of Radiative Processes   129
10.3   Quantum Mechanical Description of Gamma Decay   133
10.4   Selection Rules   135
10.5   Internal Conversion   136

CHAPTER 11
Nuclear Reactions   141

11.1   General Classification of Reactions and Conservation Laws   141
11.2   Inelastic Scattering   143
11.3   Nuclear Reactions   146
11.4   Deuteron Stripping Reactions   148
11.5   Neutron Reactions   150
11.6   Coulombic Effects   155

CHAPTER 12
Fission Reactions   158

12.1   Basic Properties of Fission Processes   158
12.2   Induced Fission   162
12.3   Fission Processes in Uranium   163
12.4   Neutron Cross Sections for Uranium   165
12.5   Critical Mass for Chain Reactions   167
12.6   Moderators and Reactor Control   168
12.7   Reactor Stability   171
12.8   Reactor Design   171

CHAPTER 13
Fusion Reactions   175

13.1   Fusion Processes   175
13.2   Fusion Cross Sections and Reaction Rates   178
13.3   Stellar Fusion Processes   181
13.4   Fusion Reactors   184
13.5   Progress in Controlled Fusion   191

Part IV   Particle Physics   197

CHAPTER 14
Particles and Interactions   199

14.1   Classification of Particles   199
14.2   Properties of Leptons   200
14.3   Feynman Diagrams   202

CHAPTER 15
The Standard Model   207

15.1   Evidence for Quarks   207
15.2   Composition of Light Hadrons   209
15.3   Composition of Heavy Hadrons   213
15.4   More About Quarks   215
15.5   Color and Gluons   218

CHAPTER 16
Particle Reactions and Decays   222

16.1   Reactions and Decays in the Context
        of the Quark Model   222
16.2 $W^\pm$ and Z^0 Bosons   227
16.3   Quark Generation Mixing   231
16.4   Conservation Laws and Vertex Rules   232
16.5   Classification of Interactions   234
16.6   Transition Probabilities and Feynman Diagrams   235
16.7   Meson Production and Fragmentation   239
16.8   CP Violation in Neutral Meson Decays   241

CHAPTER 17
Grand Unified Theories and the Solar Neutrino Problem   245

17.1   Grand Unified Theories   245
17.2   Solar Neutrinos   247
17.3   Neutrino Oscillations   252
17.4   Neutrino Masses   256

APPENDIX A
Physical Constants and Conversion Factors   259

APPENDIX B
Properties of Nuclides   261

Bibliography   277
Index   279

Preface

This text is intended for a one-semester senior-level course in nuclear and particle physics that is designed for physics majors. The course is also taught as an introductory course for graduate students who have not previously had an undergraduate course in this subject. The same material with some exclusions (primarily some of the material in Chapters 9 and 10) has been used for a junior-level course. In general, students should have a minimum background of one semester of electricity and magnetism and one semester of introductory quantum mechanics or modern physics at the sophomore level (although a semester of junior-level quantum mechanics would be desirable). The text contains about 20 percent more material that can be comfortably covered in a one-semester course (about 13–14 weeks) and gives the instructor some freedom in selecting topics in the latter half of the book.

Additional resources are available to accompany this book. A Student Solutions Manual, containing solutions to all of the even-numbered problems in the book, is available for sale to students. An Instructor's Solutions Manual, containing solutions to all of the problems, can be downloaded by qualified adoptors from the book's password-protected website at www.brookscole.com. For further information about either of these ancillaries, instructors should contact their local Cengage Learning sales representative.

I am grateful for the assistance of numerous individuals in the development and preparation of this book. Most significantly, I am indebted to the students who have taken my nuclear and particle physics course over the years. They have provided invaluable feedback on the ideas presented here and have tested countless homework problems. To David Kiang, I am grateful for the support and encouragement he has provided for my interest in teaching this course and for his numerous comments and suggestions on the manuscript. The following reviewers, who have provided numerous invaluable comments and suggestions, also deserve my thanks: Carrol Bingham, University of Tennessee; F. Paul Brady, University of California, Davis; Calvin Johnson, San Diego State University; Amitabh Lath, Rutgers University; G. E. Mitchell, North Carolina State University; Michael Schulz, University of Missouri, Rolla; and Phillip C. Womble, Western Kentucky University. I also thank Yoichiro Suzuki of the Super-Kamiokande Collaboration for providing some of the results presented in Chapter 17. Finally, I am most appreciative for the assistance provided by Ewa Dunlap, Jody O'Brien, Ariel Sibley, Stephen Penney, and David Kelly in the preparation of the text and figures.

Richard A. Dunlap

Introduction

Basic Concepts

1.1 INTRODUCTION

Much of our understanding of physics comes from the application of funda-mental theoretical concepts to the description of physical systems. This is par-ticularly true in the fields of atomic physics and solid state physics. An important example is the calculation of the energy levels of the electron in a hydrogen atom (see Chapter 5 for a more detailed discussion), where the three-dimensional time independent Schrödinger equation,

$$-\frac{\hbar^2}{2m}\nabla^2\psi + V(r)\psi = E\psi, \tag{1.1}$$

is solved. Here $\hbar$ is Planck's constant divided by 2π, m is the electron mass, $V(r)$ is the potential energy, ψ is the electron wave function, and E is the energy. Using the appropriate form of the Coulomb potential for $V(r)$, the energy eigenvalues can be easily obtained. The success of this approach in predicting the experi-mentally observed values, relies on a good fundamental understanding of the electromagnetic interactions responsible for the determination of the potential. The difficulty in applying equation (1.1) to atoms with many electrons lies, not in our lack of understanding of the fundamental properties of the electromag-netic interaction, but in the complexity of the mathematics. The situation in solid state physics is very similar. However, in nuclear physics the form of $V(r)$ has not been uniquely determined and a phenomenological approach is usually adopted. Although considerable success has been achieved by using potentials

3

based on empirical observation and on what seems sensible, there is much of a fundamental nature concerning nuclear interactions that still needs to be learned. A better appreciation of nuclear physics can be gained by some understanding of the interactions between particles that are responsible for determining nuclear properties. It is with this idea in mind that the first two chapters of this book cover, not only a discussion of some basic concepts in nuclear physics, but a brief overview of particle properties and interactions. This provides a more fundamental basis for models that will be discussed in Part II of the book.

1.2 TERMINOLOGY AND DEFINITIONS

An atom consists of a nucleus and the atomic electrons bound to it by the coulombic interaction. The simplest nucleus, that of the ^{1}H atom, consists of a single proton. All other nuclei are bound systems consisting of some combination of Z protons and N neutrons. It is the purpose of Parts II and III of this book to describe the properties of such nuclei. Such properties include the size, shape, mass, stability, and electromagnetic moments of the nucleus as well as its cross section for various reactions. This chapter covers some basic background materials that will be helpful before proceeding to Part II of the text. We begin with some basic definitions that are used in nuclear physics. Following these, we discuss the units that are commonly used to describe nuclear quantities and some sources for further information.

Atomic number (Z) is the number of protons in the nucleus of an atom and also the number of electrons in a neutral atom.

Nucleon is the general term used to refer to a neutron or a proton when it is part of a nucleus.

Atomic mass unit (u) is a unit of mass used for atoms, nuclei, or particles that is one-twelfth of the mass of a neutral atom of ^{12}C (six protons, six neutrons, and six electrons). The superscript in front of the elemental name indicates the total number of nucleons.

Atomic mass is the mass of a neutral atom (usually expressed in atomic mass units) and includes the masses of the protons, neutrons, and electrons as well as all binding energies.

Nuclear mass is the mass of the nucleus (usually expressed in atomic mass units) and includes the masses of the protons and neutrons as well as the nuclear binding energy, but does not include the mass of the atomic electrons or electronic binding energy.

Mass number (A) is the total number of nucleons in a nucleus ($A = Z + N$) and is also the integer closest to the nuclear mass in atomic mass units.

Nuclide is a specific nuclear species specified by the values of N and Z.

Isotopes are members of a family of nuclides with a common value of Z.

Radioisotopes are members of a family of unstable nuclides with a common value of Z.

Isotones are members of a family of nuclides with a common value of N.

Isobars are members of a family of nuclides with a common value of A.

Natural abundance is the relative proportion (as a fraction or as percent) of a particular isotope of an element in a naturally occurring terrestrial sample of the element.

Atomic weight is the average atomic mass of naturally occurring isotopes of an element weighted by their natural abundance.

1.3 UNITS AND DIMENSIONS

In general the units used in this book are those that are in common use in the fields of nuclear and particle physics. In most cases these are derived from standard SI units, although a few comments are necessary here to define units that are particular to these fields. The quantities that are most commonly encountered are distance, area, time, energy, and mass. These are considered below.

Distance Nuclear dimensions, as we will see in Chapter 3, are of the order of 10^{-15} m. This represents one femtometer and is abbreviated fm. In nuclear physics this unit is commonly called a fermi.

Area In SI, area units are multiples of m^2 and are most commonly encountered when discussing cross sections. Nuclear cross sections vary over a wide range of values but are frequently in the range of hundreds of fm^2. The unit barn is defined as $100\ fm^2$.

Time The time scale for nuclear processes is usually quite short. However, lifetimes for nuclear or particle decays can cover a very large range of values, from less than 10^{-25} s to more than 10^{39} s. In many cases long time periods may be expressed in years. Intermediate time periods may be expressed in minutes, hours, or days as appropriate.

Energy Energy is most commonly expressed in multiples of electron volts (eV), usually MeV (10^6 eV) in nuclear physics or GeV (10^9 eV) in particle physics. In some cases keV is also used when energies are in this range. The eV is related to the conventional SI energy unit (the Joule) by $1\ eV = 1.602 \times 10^{-19}$ J.

Mass Atomic and nuclear masses are most commonly given in atomic mass units (u) as defined above. This unit is related to the conventional SI mass unit (the kilogram) as $1\ u = 1.661 \times 10^{-27}$ kg. Masses of subatomic particles are sometimes given in atomic mass units, particularly when these masses are used in the context of nuclear properties. When dealing with the properties of the particles themselves, it is more common to express masses in units of MeV/c^2 or GeV/c^2 (or some multiple of these quantities). These mass units result from Einstein's equivalence of mass and energy expressed as

$$E = mc^2. \tag{1.2}$$

The conversion between atomic mass units and MeV/c^2 is given by: 1 u = 931.502 MeV/c^2.

1.4 SOURCES OF INFORMATION

There are numerous textbooks on nuclear and particle physics that can provide supplemental information. A list of recommended books appears in the Bibliography. There are a number of sources for nuclear data such as atomic masses, decay properties, reaction cross sections, and particle masses. The *Table of Isotopes*, 8th edition, edited by R. B. Firestone and V. S. Shirley (Wiley: New York, 1998) is the standard reference for information about nuclear decays and is now available both in book and CD formats. There are a number of sites on the Internet that provide useful nuclear data. Some that are particularly useful are as follows:

Japan Atomic Energy Research Institute (http://wwwndc.tokai.jaeri.go.jp) has very convenient tables of nuclear data that include atomic masses for the nuclides.

Web Elements (http://www.webelements.com/) is aimed at providing chemical and physical data for the elements and includes a number of important nuclear properties as well.

Table of Nuclides (http://www2.bnl.gov/ton/) provides much of the information that is available in the *Table of Isotopes* book. Some decay information may not be as complete as that given in the book.

Brookhaven National Laboratory, National Nuclear Data Center (http://www.nndc.bnl.gov) gives nuclear data as well as links to useful computer routines for nuclear physics.

Particle Data Group (http://pdg.lbl.gov) provides up-to-date information about subatomic particles and includes all fundamental properties such as mass, charge, and spin.

Particles and Interactions

2.1 CLASSIFICATION OF SUBATOMIC PARTICLES

Our initial discussion of nuclear properties will deal almost exclusively with neutrons and protons. In Part III of the text we will encounter electrons and neutrinos and in Part IV we will examine the properties of a wide variety of subatomic particles. Although the details of the properties of subatomic particles, including the neutron and the proton, will be given in the latter portions of the text, a brief overview of the classification and some of the properties of these particles will be presented in this chapter to allow for a more complete discussion of the interactions that take place within the nucleus.

All subatomic particles can be categorized as either fermions or bosons depending on whether their spin is $\frac{1}{2}$ integral or integral, respectively. The former particles can be described by Fermi-Dirac statistics and the latter by Bose-Einstein statistics. The general division of fermions and bosons into categories based on other particle properties is summarized in Figure 2.1. It is seen from the figure that the fermions are either leptons or baryons. The bosons are categorized as either mesons or gauge bosons. As will be demonstrated in Chapter 15, baryons and mesons share certain properties and are collectively referred to as *hadrons*. The basic properties of the particles of relevance to Parts II and III of this book are summarized in Table 2.1.

The properties of the nucleus depend on the properties of these particles as determined by the interactions that act on each class of particle and by the

Table 2.1 | Properties of Some Subatomic Particles That Are of Importance in Nuclear Physics

Particle	Symbol	Classification	Lepton Number	Baryon Number	Charge (e)	Mass (MeV/c^2)	Lifetime (s)
Electron	e^-	lepton	+1	0	−1	0.511	∞
Positron	e^+	lepton	−1	0	+1	0.511	∞
Electron neutrino	ν_e	lepton	+1	0	0	≈0	∞
Anti-electron neutrino	$\overline{\nu}_e$	lepton	−1	0	0	≈0	∞
Proton	p	baryon	0	+1	+1	938.28	>10^{39}
Neutron	n	baryon	0	+1	0	939.57	898

conservation laws that are applicable in each case. A selective interaction is, for example, the electrostatic interaction that acts only on objects that are electrically charged. An example of a conservation law is the conservation of charge; a positively charged object and a negatively charged object may interact but the total charge (sum of the positive and negative charges) will be the same before and after the interaction.

The remainder of this chapter gives a brief overview of the properties of the various classes of particles that are of relevance to the properties of the nucleus. This includes a discussion of the interactions that apply to each type of particle and the conservation laws that must be considered.

Figure 2.1 | Classification of Fermions and Bosons

This diagram shows some examples of particles in each category.

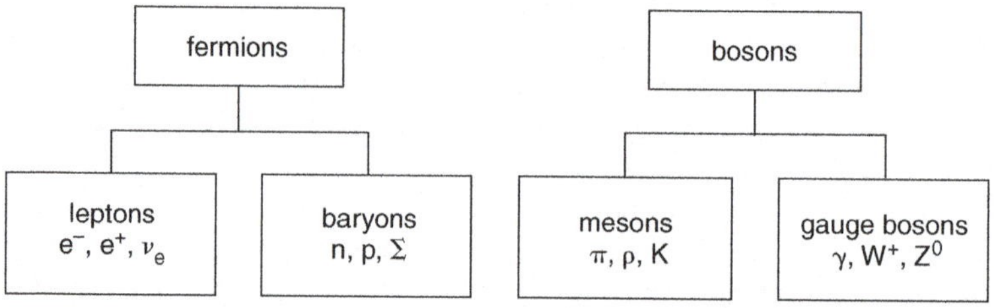

Table 2.2	Properties of the Four Interactions in Nature. The Relative Strength Is Normalized to Unity for the Strong Interaction

Interaction	Acts on	Strength	Range
Strong	hadrons	1	10^{-15} m
Electromagnetic	electric charges	10^{-2}	long $(1/r^2)$
Weak	leptons and hadrons	10^{-5}	10^{-18} m
Gravity	masses	10^{-39}	long $(1/r^2)$

2.2 CLASSIFICATION AND RANGES OF INTERACTIONS

There are four known forces in nature as summarized in Table 2.2. The table gives the relative strengths of these interactions and the range over which they act. The gravitational interaction is sufficiently weak that it plays no role in the behavior of nuclei. The electromagnetic interaction is long range and affects macroscopic objects as well as nuclei; the strong and weak interactions are short range and are only important on a size scale comparable to the dimension of the nucleus. The electromagnetic interaction acts on objects that possess charge. Along similar lines the strong and weak interactions each act on certain types of particles. Although the weak interaction is very weak and very short range (compared with the strong interaction), it is of great importance to the behavior of nuclei because it acts upon leptons as well as hadrons. The strong interaction does not act on leptons so the weak interaction is an important consideration for many processes that involve only leptons. In fact, Chapters 15 and 16 will show that strong and weak interactions act differently on hadrons and that the weak interaction is significant even in many processes that involve only hadrons.

2.3 CONSERVATION LAWS

A detailed analysis of any nuclear process requires a consideration of the applicable conservation laws. Certain quantities (for example, angular momentum) are believed to be conserved in all processes while other quantities (for example, parity) may or may not be conserved depending on the nature of the interactions that are present. Table 2.3 summarizes the quantities that are conserved for the strong, electromagnetic, and weak interactions. An important distinction between the strong interaction and the weak interaction is seen from the table. Certain quantities (for example, isospin, strangeness, etc.) are not conserved in processes that are dominated by the weak interaction while these same quantities must be conserved in processes involving only the strong interaction. Details of these quantities will be discussed in Part IV of the book. When the weak

Table 2.3 | Quantities That Are Conserved (Y) and NonConserved (N) for the Strong, Electromagnetic, and Weak Interactions

| | Interaction | | |
Quantity	Strong	Electromagnetic	Weak
Mass/energy	Y	Y	Y
Linear momentum	Y	Y	Y
Angular momentum	Y	Y	Y
Charge	Y	Y	Y
Parity	Y	Y	N
Lepton number	Y	Y	Y
Baryon number	Y	Y	Y
Lepton generation	Y	Y	Y
Isospin	Y	N	N
Strangeness	Y	Y	N
Charm	Y	Y	N
Bottom	Y	Y	N
Top	Y	Y	N

interaction is present, it is the non-conservation of these quantities that allows certain processes (such as β-decay) to occur.

Nuclear processes must satisfy all appropriate conservation laws. Mass/energy and angular momentum must be conserved in all nuclear processes as is the case for macroscopic processes. Of particular relevance to nuclear processes is the conservation of charge, lepton number, and baryon number. These quantities are shown for particles of interest in Table 2.1. An investigation of reactions and decay processes as described in Parts II and III of this book will demonstrate the conservation of these quantities. Further discussion on the conservation or nonconservation of certain quantities in reactions and decays of particles will be presented in Part IV.

Nuclear Properties and Models

Nuclear Composition and Size

3.1 COMPOSITION OF THE NUCLEUS

The physical properties of a nucleus are determined by its configuration of nucleons. As discussed in Chapter 5, the nucleon configuration is determined by the quantum numbers (spin, energy, etc.) associated with each of these particles. The most important factor in determining nuclear stability is the number of neutrons and protons in the nucleus. The behavior of a system of neutral neutrons and positively charged protons is determined by both the electromagnetic interaction and the strong interaction. In order for a nucleus to be formed, the coulombic repulsion between the protons must be overcome by an attractive strong interaction that involves both the neutrons and the protons. A certain combination of neutrons and protons may form a stable nucleus, it may form an unstable nucleus, or it may not form a nucleus at all. One might expect that adding more neutrons would help to stabilize a nucleus because it increases the attractive strong interaction but does not influence the repulsive coulombic interaction between protons. However, this is not entirely true. As we can see from the example in Table 3.1, carbon forms two stable isotopes with $Z = 6$ and $N = 6$ or $N = 7$. Fewer than six neutrons combined with six protons do not form a stable nucleus. This is, at least in part, due to an insufficient attractive strong interaction to overcome the repulsive coulombic interaction. We also see that more than seven neutrons combined with six protons does not form a stable nucleus. In fact, it can be seen from the table that the more the number of neutrons differs from six or seven the greater the nuclear instability as indicated by

Table 3.1 | Properties of the Known Isotopes of Carbon

Nuclide	Atomic Mass (u)	Stability	Natural Abundance (%)	Lifetime
^{9}C	9.03140087	unstable	0	182 ms
^{10}C	10.01685311	unstable	0	27.8 s
^{11}C	11.011433818	unstable	0	29.4 m
^{12}C	12.000000000	stable	98.9	–
^{13}C	13.003354838	stable	1.2	–
^{14}C	14.003241988	unstable	0	8267 y
^{15}C	15.010599258	unstable	0	3.533 s
^{16}C	16.014701243	unstable	0	1.078 s
^{17}C	17.022583712	unstable	0	278 ms

a decreasing mean lifetime. Thus it is clear that for a given number of protons neither too few nor too many neutrons will yield a stable nucleus. The reasons for these properties will become clear in Chapters 4 and 5.

The known nuclides, both stable and unstable as a function of N and Z are illustrated in Figure 3.1. The general features shown in this figure can be summarized as follows:

1. The unstable nuclides form a region on either side of the stable ones.
2. For small values of Z there is a tendency for $N = Z$.
3. For large values of Z, $N > Z$.

The points in Figure 3.1 are discrete since N and Z are integers. It can be shown that the stability of a nucleus is closely related to whether N and/or Z is even or odd. In fact as illustrated by the number of β-stable nuclides given in Table 3.2 some very remarkable trends can be observed (β-stability will be discussed in detail in Chapter 4). In this table nuclides are defined as even–even, even–odd, odd–even, or odd–odd depending on whether N and Z are even or odd. The four situations defined above correspond to values of A that are even, odd, odd, and even, respectively. These data clearly show that even–even nuclei have a greater probability of being stable, followed by approximately equal probabilities for even–odd and odd–even nuclei, while almost no odd–odd nuclei are stable.

The models that will be developed in Chapters 4 and 5 will describe the tendency of N to equal Z for small A and the nuclear stability as a function of an even or odd number of nucleons. However, before proceeding to develop specific models to describe nuclear properties, it is important to understand a little about the size and structure of the nucleus as discussed in the remainder of this chapter.

Table 3.2 | Number of β-Stable Nuclei for Various Combinations of N and Z

N	Z	A	Number Known
even	odd	odd	50
odd	even	odd	55
even	even	even	165
odd	odd	even	4

Figure 3.1 | Relationship Between N and Z for Known Stable (Black Symbols) and Unstable (Gray Area) Nuclides

Lines for N = Z and a typical constant value of A are shown.

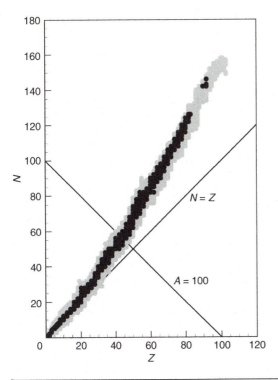

3.2 RUTHERFORD SCATTERING

Much of what has been learned about the structure of the nucleus and the distribution of matter within the nucleus has come from studies of scattering cross sections. The simplest type of scattering that can be observed is Rutherford scattering. This is the result of the nonrelativistic coulombic scattering of point charges. Deviations of experimental results from the scattering cross sections predicted by the Rutherford formula are very important in understanding the structure of the nucleus. We begin, therefore, with the basic nonrelativistic derivation of the Rutherford scattering cross section. This simple derivation assumes that the nucleus is sufficiently massive that it does not recoil during the scattering process.

The geometry of the Rutherford scattering problem is illustrated in Figure 3.2. The scattering particle (with charge ze) approaches the scatterer (with charge Ze) at an impact parameter defined in the figure as b and with an initial velocity v_0. At any time the location of the scattering particle is given relative to the scatterer by the coordinates (r, ϕ) where the angle ϕ is measured relative to the line AB. For the point $\phi = 0$ the radial velocity of the particle is zero and the distance, r, is a minimum. Conservation of angular momentum requires that

$$mv_0 b = mr^2 \frac{d\phi}{dt}.$$

(3.1)

Figure 3.2 | Geometry for the Rutherford Scattering Problem

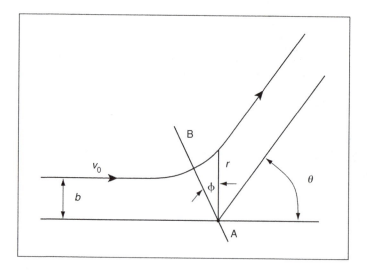

The force acting on the particle due to the coulombic interaction is given in terms of the linear momentum, p, as

$$\vec{F} = \frac{d\vec{p}}{dt} = \frac{Zze^2}{4\pi\varepsilon_0 r^3}\vec{r}. \tag{3.2}$$

The change in momentum over a time interval t_1 to t_2 is found by integrating (3.2);

$$\Delta\vec{p} = \int_{t_1}^{t_2} \vec{F}\,dt. \tag{3.3}$$

The time at which the particle is at $\phi = 0$ is defined as $t = 0$. The linear momentum of the particle at $t = -\infty$ along the direction AB is

$$p = -mv_0\sin\frac{\theta}{2} \tag{3.4}$$

where θ is the scattering angle as defined in the figure. Over the time interval $t = -\infty$ to $+\infty$, equation (3.3) can be written in terms of the Coulomb force as

$$2mv_0\sin\frac{\theta}{2} = \int_{-\infty}^{+\infty} \frac{Zze^2}{4\pi\varepsilon_0 r^2}\cos\phi\,dt. \tag{3.5}$$

A change of variables in the integration from t to ϕ can be accomplished using equation (3.1) to give

$$\sin\frac{\theta}{2} = \frac{Zze^2}{8\pi\varepsilon_0 mv_0^2 b} \int_{-(\pi-\theta)/2}^{+(\pi-\theta)/2} \cos\phi\,d\phi. \tag{3.6}$$

Integration of equation (3.6) is straightforward and yields

$$\tan\frac{\theta}{2} = \frac{Zze^2}{4\pi\varepsilon_0 mv_0^2 b}. \tag{3.7}$$

This expression shows that increasing the impact parameter decreases the scattering angle. Thus all particles that are scattered by an angle greater than some value of θ, must have impact parameters less than some value of b. Incident particles with b less than a particular value define an area, or cross section, given by

$$\sigma = \pi b^2. \tag{3.8}$$

Thus it can be said from equations (3.6) and (3.7) that the scattering cross section for scattering by an angle greater than θ is

$$\sigma = \pi\left[\frac{Zze^2\cot\dfrac{\theta}{2}}{4\pi\varepsilon_0 mv_0^2}\right]^2. \tag{3.9}$$

Figure 3.3 | Relationship Between $d\theta$ and $d\Omega$ for the Rutherford Scattering Problem

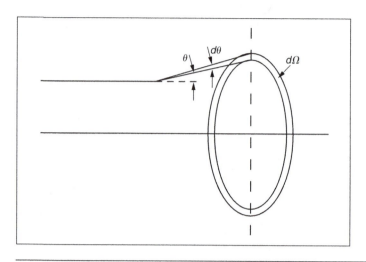

Experiments typically measure the differential scattering cross section, that is the scattering into a differential solid angle, $d\Omega$. Some simple solid geometry gives

$$d\Omega = 2\pi \sin\theta d\theta \qquad (3.10)$$

where the differential angle as shown in Figure 3.3 has been integrated to account for the axial symmetry of the problem. The differential scattering cross section is defined as

$$\frac{d\sigma}{d\Omega} = \frac{d\sigma}{d\theta}\frac{d\theta}{d\Omega} \qquad (3.11)$$

or from equations (3.9) and (3.10)

$$\frac{d\sigma}{d\Omega} = \left[\frac{Zze^2}{8\pi\varepsilon_0 mv_0^2}\right]^2 \csc^4\frac{\theta}{2}. \qquad (3.12)$$

This is known as the Rutherford differential scattering cross section and will be used in the next section for the analysis of nuclear charge distributions. Although our analysis was based entirely on classical arguments, it is a fortunate coincidence that a thorough quantum mechanical treatment yields the exact same answer. This is true for all scattering problems involving a $1/r$ potential. A final note of interest here concerns the sign of the coulombic interaction. Throughout this derivation it has been implied that this is repulsive as indicated by Figure 3.2. Because equation (3.12) shows that the cross section is proportional to $(Zz)^2$ the

Figure 3.4 | Particle Trajectories for Rutherford Scattering for (a) Repulsive Interaction and (b) Attractive Interaction

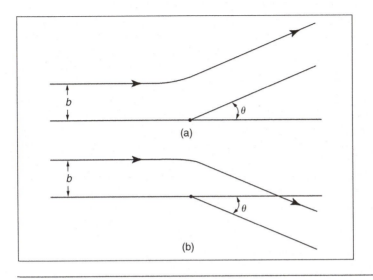

sign of Zz is irrelevant. Figure 3.4 shows that although the trajectory of the scattered particle is not exactly the same for repulsive and attractive scattering, the relationship between impact parameter b and scattering angle, θ, is independent of the sign of the interaction.

3.3 CHARGE DISTRIBUTION OF THE NUCLEUS

The distribution of charge in the nucleus is determined by the distribution of protons, $\rho_p(r)$. Initial scattering experiments conducted by Rutherford to study the structure of the atom used the scattering of α-particles (^{4}He nuclei) and observed the general size of the nucleus. Since the α-particles are nuclei and have their own structure, it is more common now to utilize electron scattering to observe these phenomena. In order to properly analyze the results of scattering experiments it is important to include two factors that have, thus far, been omitted—relativistic effects and the spatial distribution of charges within the nucleus. Relativistic effects can be included in the derivation given above and leads to a relativistic differential scattering cross section given by

$$\left(\frac{d\sigma}{d\Omega}\right)_{rel} = \left(\frac{d\sigma}{d\Omega}\right)_{nonrel}\left[1 - \frac{v_0^2}{c^2}\sin^2\frac{\theta}{2}\right]. \tag{3.13}$$

Inclusion of the nonpoint charge nature of the nucleus is a substantially more complex problem but leads to some highly significant information about nuclear forces and structure. In general the effect of a finite size nucleus is to reduce the scattered intensity at all angles (except zero) from the prediction for scattering of point charges. The reasons for this can be seen by drawing an analogy with the diffraction of light waves by a circular aperture. For any scattering angle the total measured intensity is the result of interference effects between different components of the beam that are diffracted from different points within the aperture. This gives rise to the characteristic ring-like pattern. Similarly for a beam of, say, electrons scattered from a finite size nucleus the interference (of the wave functions) of electrons scattered from different locations within the nuclear charge distribution gives rise to oscillations in the scattered intensity as a function of scattering angle. The nature of these oscillations depends on the wavelength—that is, energy—of the electrons as well as the spatial distribution of charges within the nucleus. The former quantity is known and can be controlled in the experiment. The latter quantity can be extracted from the experimental measurement of scattered intensity as a function of scattering angle. The above description can be quantified by writing the differential scattering cross section as

$$\frac{d\sigma}{d\Omega} = \left(\frac{d\sigma}{d\Omega}\right)_{Rutherford} |F(\theta)|^2 \tag{3.14}$$

where the form factor $F(\theta)$ is a function that is characteristic of the nuclear charge distribution $\rho_p(r)$ but also depends on the energy of the scattering particles. Here we have assumed that the charge distribution of the nucleus is spherically symmetric. This is a reasonable approximation for most, but not all, nuclei (see Chapter 6). The form factor can be related to $\rho_p(r)$ as

$$F(\theta) = \frac{1}{Ze} \int \rho_p(\vec{r}) e^{i\delta(\theta,\vec{r})} dV \tag{3.15}$$

where the integration is over the volume of the nucleus and δ is a phase factor given by

$$\delta = \frac{\Delta\vec{p} \cdot \vec{r}}{\hbar}. \tag{3.16}$$

The magnitude of the change in the momentum is given by (3.4) as

$$|\Delta\vec{p}| = 2p\sin\frac{\theta}{2}. \tag{3.17}$$

Knowing $\rho_p(r)$ allows for the calculation of $F(\theta)$ as illustrated for a simple case in Figure 3.5. This demonstrates the expected oscillatory nature of the form factor as a function of angle with $F(\theta)$ normalized to unity for $\theta = 0$. From an experimental standpoint, however, $\rho_p(r)$ is not known a priori and equation (3.15) may be substituted into (3.14) in order to relate $\rho_p(r)$ to the measured differential scattering cross section. Two approaches can be taken to analyze experimental data on the basis of this integral: a model independent numerical

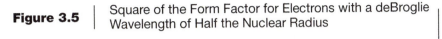

Figure 3.5 | Square of the Form Factor for Electrons with a deBroglie Wavelength of Half the Nuclear Radius

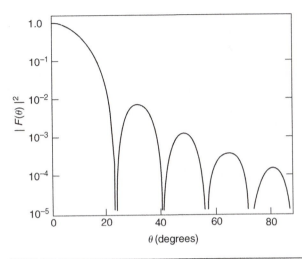

technique to obtain $\rho_p(r)$ and the assumption of a particular functional form for the radial dependence of $\rho_p(r)$ and the fitting of the experimental data to the above equations to obtain the numerical values of unknown parameters. The latter is not necessary for the analysis of scattering data but provides a convenient analytical expression for the nuclear charge distribution that can be implemented in nuclear model calculations. We will begin with a discussion of the former method since it is more general.

Figure 3.6 shows some typical experimental data for scattering of electrons from a nucleus. This clearly shows the combined effects of the geometric factor predicted by the Rutherford formula and the oscillatory nature introduced by the form factor. Figure 3.7 shows the charge distribution that is extracted from these data. In general, such analyses of scattering data show similar features; a relatively flat region near the center and a gradual decrease in density towards the edge. A good representation of this shape is provided by the Woods-Saxon model:

$$\rho_p(r) = \frac{\rho_{p0}}{1 + \exp[(r - R_0)/a]}. \tag{3.18}$$

R_0 gives the mean nuclear radius (the radius at which the density drops to $1/2$ of its maximum value) and a is related to the width of the edge region. The coefficient ρ_{p0} in this function is given by normalizing the total integrated nuclear charge as

$$Z = \int \rho_p(r) dV. \tag{3.19}$$

Figure 3.6 | Measured Differential Scattering Cross Section for 450 MeV Electrons Incident on ^{58}Ni

The solid line is a fit using the charge distribution shown in Figure 3.7.

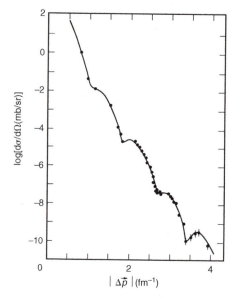

Figure 3.7 | Charge Distribution for ^{58}Ni Extracted Numerically from the Measurements Shown in Figure 3.6

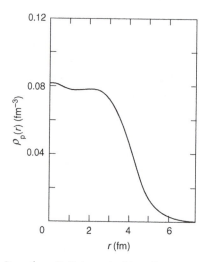

Figure 3.8 | Woods-Saxon Charge Distributions for Some Nuclei

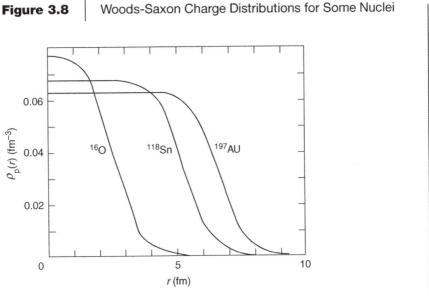

An analysis of $\rho_p(r)$ for several nuclei is shown in Figure 3.8. It is seen that the shape of this function is a reasonable approximation of the results of the model independent analysis shown above. The figure shows some important aspects of the nuclear charge distributions.

1. Larger nuclei have a larger mean diameter.
2. The edge region has a similar width in all nuclei.
3. The charge density at the center is greater in light nuclei than in heavy nuclei.

3.4 MASS DISTRIBUTION OF THE NUCLEUS

The measurement of $\rho_p(r)$ is based on the distribution of charge and hence the distribution of protons in the nucleus. It is also important to know the distribution of all nucleons (neutrons and protons) in the nucleus. Since neutrons are uncharged, coulombic scattering experiments do not observe them. It is, however, fairly straightforward to infer the total nucleon density (or nuclear mass density), $\rho(r)$, from measurements of $\rho_p(r)$. A nucleus with N neutrons and Z protons has an overall neutron to proton ratio N/Z, and a total number of nucleons $A = N + Z$. It is usually assumed that the ratio of neutron density, $\rho_n(r)$, to proton density, $\rho_p(r)$, is the same everywhere within the nucleus. This is written as

$$\frac{\rho_n(r)}{\rho_p(r)} = \frac{N}{Z}. \tag{3.20}$$

Figure 3.9 | Woods-Saxon Mass Distributions for Some Nuclei

Note the differences between these distributions and the charge distributions for the same nuclei shown in Figure 3.8.

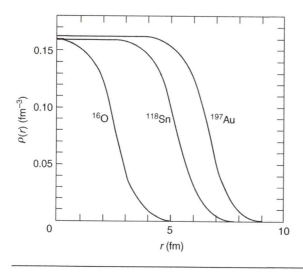

It is also known that the total density is the sum of the neutron and proton densities

$$\rho(r) = \rho_p(r) + \rho_n(r).$$ (3.21)

These two equations can be easily solved to give

$$\rho(r) = \rho_p(r)\left[1 + \frac{N}{Z}\right].$$ (3.22)

Using this equation, data of the form shown in Figure 3.8 can be rescaled for different N and Z to give the total nucleon density as illustrated in Figure 3.9. It is significant here that the mean nuclear radii and the width of the edge regions has not changed from that shown in Figure 3.8. It is also seen that all nuclei, regardless of total mass, have virtually the same nucleon density in their centers. This is a very important feature of nuclei and confirms that the strong interaction is very short ranged and is dominated by the interaction between nearest neighbor nucleons. This will be discussed further in the next chapter. The constant a in the Woods-Saxon equation is found to be virtually independent of total nuclear mass and has a value of 0.52 ± 0.01 fm. The mean nuclear radius, R_0, increases with total nuclear mass and an analysis of the experimental results leads to the empirical expression

$$R_0 = 1.2 \times A^{1/3} \text{ (fm)}.$$ (3.23)

This is what would be anticipated for nuclei with differing mass but the same density. This will play an important role in the development of the liquid drop model described in Chapter 4.

Problems

3.1. (a) For the coulombic scattering of a point charge $+ze$ from a heavy nucleus of charge $+Ze$, calculate the distance of closest approach for $b = 0$. Ignore the recoil of the heavy nucleus.

(b) For $b = 0$ calculate the point of closest approach for an 8 MeV α-particle incident on an Au nucleus and compare with the expected nuclear radius.

3.2. For a nonrelativistic coulombic scattering experiment the number of particles scattered per unit time at a fixed scattering angle can be measured as a function of incident particle energy. Describe the expected results of such an experiment.

3.3. α-particles (^{4}He nuclei) are incident on Au nuclei. What is the minimum α-particle energy necessary for the two nuclei to come in contact?

3.4. (a) For an 8 MeV α-particle incident on an Au nuclei, what is the impact parameter when the particle is scattered at 90°?

(b) What is the point of closest approach for the α-particle scattered at 90°?

(c) What is the kinetic energy of the α-particle at the point of closest approach?

3.5. Calculate the density of matter at the center of a ^{208}Pb nucleus.

3.6. (a) Using the data shown in Figure 3.9, estimate the width of the surface region of a nucleus; that is, the distance over which the density drops from 90% of its central value to 10% of its central value.

(b) Using the result of part (a), estimate the value of a in equation (3.18).

3.7. (a) For 10 MeV α-particles incident on Au nuclei, calculate the total scattering cross section for scattering angles $\theta > 1°$, $\theta > 5°$, and $\theta > 20°$.

(b) For the conditions given in part (a), calculate the differential scattering cross section for $\theta = 1°$, $\theta = 5°$, and $\theta = 20°$.

3.8. (a) For the scattering of 0.1 MeV electrons from ^{119}Sn nuclei, calculate the relative size of the relativistic correction to the differential scattering cross section for scattering angles of 20° and 90°.

(b) Repeat part (a) for 1 MeV and 100 MeV electrons.

3.9. (a) Using Figure 3.9, estimate the fraction of nucleons that are in the surface region of an ^{16}O nucleus.

(b) Repeat part (a) for ^{118}Sn and ^{197}Au nuclei.

3.10. (a) For a 10 MeV α-particle incident on an ^{16}O nucleus with an impact parameter $b = 0$, calculate the recoil energy of the nucleus.

(b) Repeat part (a) for 10 MeV α-particles incident on ^{118}Sn and ^{197}Au nuclei.

4 CHAPTER | Binding Energy and the Liquid Drop Model

4.1 DEFINITION AND PROPERTIES OF THE NUCLEAR BINDING ENERGY

The nuclear binding energy, B, is the minimum energy required to separate a nucleus into its component neutrons and protons. The mass equivalent of the nuclear binding energy relates the nuclear mass (m_N) to the mass of the component neutrons (m_n) and protons (m_p):

$$m_N = Nm_n + Zm_p - \frac{B}{c^2}.$$

(4.1)

B is defined to be explicitly positive for a bound system, so from equation (4.1) it is seen that the binding energy of the nucleus decreases its total mass. For a given contribution to the mass from the neutrons and protons, the nuclear stability is increased as B increases, and this corresponds to a reduction in total mass. Nuclear masses are usually not measured directly. Instead, for a given nuclide the atomic mass of a neutral atom is generally measured. This can be related to the nuclear mass by

$$m = Nm_n + Z(m_p + m_e) - \frac{B}{c^2} - \frac{b}{c^2}$$

(4.2)

where b is the total binding energy associated with all the atomic electrons. In almost all cases b is sufficiently small that it can be ignored in calculations involving nuclear masses and binding energies. Since calculations of decay energies involve differences in atomic masses the effects of the electronic binding energy tend to cancel out in any case. However, in cases where it is necessary to include this term in equation (4.2), it is often sufficient to use the results of a semiclassical calculation that gives

$$b = 20.8 \times Z^{7/3} \quad (eV). \tag{4.3}$$

Atomic masses are usually given in atomic mass units and are tabulated in Appendix B. Tables in some references give the mass excess Δ (usually in MeV), which is defined in terms of the atomic mass m as

$$\Delta = (m - Au)c^2. \tag{4.4}$$

The total nuclear binding energy shows an approximately linear relationship to A. This means, that to first order, each nucleon is bound to the nucleus with the same energy regardless of the size of the nucleus. This is consistent with the conclusions that can be drawn concerning the range of the strong interaction on the basis of nucleon density results as discussed in Chapter 3. However, on close inspection the binding energy shows some interesting features. These are most easily seen in a plot of B/A as a function of A as shown in Figure 4.1. If B is linear in A then this quantity would be a constant. The general behavior as seen in the figure can be explained by the liquid drop model. This yields the so-called semiempirical mass formula as discussed in the following section.

Figure 4.1 | Binding Energy per Nucleon as a Function of Mass Number, A

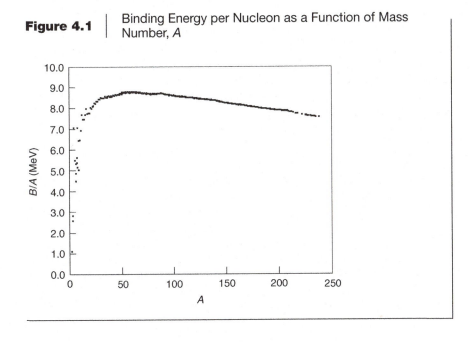

4.2 THE LIQUID DROP MODEL

Much is known about nucleon–nucleon interactions from an experimental standpoint. Evidence indicates that this interaction is both strong and short range. It is also known that nucleon–nucleon interactions are spin dependent and that they have both central and noncentral components. Although a detailed description of these interactions is complex, much can be learned about their effects on the behavior of nuclei on the basis of some simple models. Two of these, the liquid drop model, which draws analogies with the behavior of liquids, and the shell model, which is based on the quantum mechanical behavior of neutrons and protons, are discussed in detail in this book.

The liquid drop model is a semiempirical model that describes the behavior of the nuclear binding energy as a function of the number of neutrons and protons. It is also an effective method of investigating the relative stability of various nuclides. This model is based, in part, on an analogy with the properties of a drop of liquid, in part on a consideration of fundamental electrostatics and in part on empirical observations. The analogy with a liquid drop is expected to be, at least to a point, valid, as the molecular interactions (van der Waals interactions) holding a drop of liquid together are short ranged. However, we should not be tempted to extend this analogy to the point of trying to understand any of the basic properties of the strong interaction by comparison with the interactions present in a liquid. The binding energy as predicted by the liquid drop model can be expressed as the sum of a number of terms due to different contributions. These are each discussed in detail below. It is assumed here that the nucleus is spherically symmetric, which in most cases is a good approximation, and that the nuclear volume is directly related to the number of nucleons present.

Volume Term   This term results from the interaction of each nucleon with its nearest neighbor nucleons. We would expect this to be the dominant contribution to the overall interaction energy as the nucleon–nucleon interaction is very short ranged. Because the nucleus contains A nucleons (we do not distinguish between neutrons and protons here) the associated binding energy, which is proportional to the nuclear volume, is given by

$$B_V = a_V A. \tag{4.5}$$

The coefficient a_V is a constant that can be determined by comparison with experimental data. The form of this term is consistent with the short-range nature of the strong interaction. In the case of long-range interactions, such as coulombic interactions, then each nucleon would interact with each other nucleon and the resulting energy would be proportional to $A(A - 1)$; see the discussion of coulombic interactions below. In general the form given by equation (4.5) is an overestimate of the binding energy since the other interactions that will be described below (except the pairing interaction) tend to destabilize the nucleus and, therefore, decrease the total binding energy.

Surface Term   All nucleons in the nucleus are not equivalent. Those on the surface are surrounded by a smaller number of nearest neighbor nucleons than those in the interior. This means that the nucleons on the surface are less tightly bound and the binding energy given above must be decreased by a quantity that is proportional to the number of surface nucleons or to the surface area of the nucleus. For a spherical nucleus the surface area is related to the volume as $S \propto V^{2/3}$ so this contribution to the energy can be written

$$B_S = -a_S A^{2/3} \tag{4.6}$$

where the minus sign indicates that this decreases the overall nuclear stability. Again the coefficient is determined by a comparison with experimental data.

Coulomb Term   The repulsive coulombic interaction between the protons in the nucleus reduces the binding energy further. Basic electrostatics shows that the potential energy associated with a uniform spherical distribution of Z charges is proportional to $Z(Z-1)/r$, and using equation (3.23) this can be related to a term in the binding energy of the form

$$B_C = -a_C \frac{Z(Z-1)}{A^{1/3}}. \tag{4.7}$$

The coefficient a_C can be calculated analytically and has the value of 0.72 MeV. In the present discussion we will leave this as a free parameter in the analysis. One (of several) valid reasons for this approach is the nonuniform density of charges near the edge of the nucleus as was shown in Figure 3.9.

Symmetry Term   As was illustrated in Figure 3.1 there is a tendency for stable nuclei with small A to have $N = Z$. For large A, N tends to be somewhat greater than Z. On the basis of the first two terms discussed above there is no restriction placed on the relative number of protons and neutrons in the nucleus. The Coulomb term would favor nuclei with many neutrons and few protons. This is clearly not what is observed. The symmetry term favors $N = Z$, but becomes less important as A becomes large. The usual form of this term is

$$B_{sym} = -a_{sym} \frac{(A - 2Z)^2}{A} \tag{4.8}$$

where again a_{sym} is a free parameter in the analysis. Here it is important to note that $A - 2Z = N - Z$ and is a measure of the deviation from the $N = Z$ line previously shown in Figure 3.1.

Pairing Term   The behavior that was described in Table 3.2 is taken into account by the pairing term, which is of the form

$$B_p = -\frac{a_p}{A^{3/4}}. \tag{4.9}$$

Figure 4.2 | Relative Importance of the Various Contributions to the Binding Energy Predicted by the Semiempirical Mass Formula

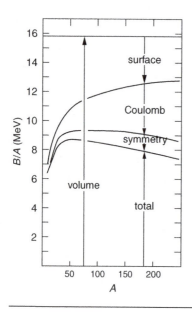

The free parameter, a_p, takes on a positive value for odd–odd nuclei indicating a decrease in stability, a negative value of even–even nuclei indicating an increase in stability and zero for odd A nuclei.

The total binding energy per nucleon as determined by the sum of the above terms and per nucleon is expressed

$$\frac{B}{A} = a_V - \frac{a_S}{A^{1/3}} - \frac{a_C Z(Z-1)}{A^{4/3}} - \frac{a_{sym}(A-2Z)^2}{A^2} - \frac{a_p}{A^{7/4}}. \qquad (4.10)$$

This can be compared with the experimental results as indicated in Figure 4.1 to obtain best fit values of the coefficients. The general shape of the curve is determined by the first four terms above. The relative importance of these terms is illustrated in Figure 4.2. The value of a_p is determined from a consideration of fluctuations in B/A as A changes from odd to even. This analysis gives the best fit values as

$$a_V = 15.5 \text{ MeV}$$

$$a_S = 16.8 \text{ MeV}$$

$$a_C = 0.72 \text{ MeV}$$

$$a_{sym} = 23.2 \text{ MeV}$$

$$a_p = +34 \text{ MeV} \qquad N, Z = \text{odd–odd}$$

$$0 \text{ MeV} \qquad A = \text{odd}$$

$$-34 \text{ MeV} \qquad N, Z = \text{even–even}. \qquad (4.11)$$

Note that the value for a_C is in agreement with the analytical result calculated for a uniform charge distribution. As discussed in the next section, these values can be used to describe the stability of certain nuclides.

4.3 BETA STABILITY

From the development given above the total atomic mass can be determined to be

$$m = (A - Z)m_n + Z(m_p + m_e) - \frac{a_V A}{c^2} + \frac{a_S A^{2/3}}{c^2} + \frac{a_C Z(Z-1)}{A^{1/3} c^2}$$

$$+ \frac{a_{sym}(A - 2Z)^2}{A c^2} + \frac{a_p}{A^{3/4} c^2} \qquad (4.12)$$

where N is written as $A - Z$ and the electronic binding energy is ignored. It is interesting to consider the Z dependence of this expression for a constant value of A. This is equivalent to examining the behavior of isobars. Collecting together powers of Z, it is seen that the above expression is a quadratic. The simplest case to consider is for nuclei with odd A since the pairing term is zero. An example of plotting m versus Z at constant A is shown in Figure 4.3 for $A = 135$.

Figure 4.3 | Mass Parabola for $A = 135$ Showing One Stable Nuclide (as Expected for Odd A) with $Z = 56$

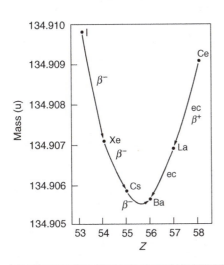

The data plotted here are experimentally measured atomic masses, but the parabolic shape of the curve as predicted by the semiempirical mass formula is evident. Minimizing the mass corresponds to maximizing nuclear stability. It is seen in the figure that the nuclide closest to the minimum in the figure will be stable; in this case it is ^{135}Ba with 56 protons. Nuclides with $A = 135$ and more or fewer than 56 protons are unstable and decay towards $Z = 56$. For a given (odd) A the minimum in the mass parabola can be calculated on the basis of the semiempirical mass formula. Setting $dm/dZ = 0$ equation (4.12) gives the value of Z, which minimizes the mass as

$$Z_{min} = \frac{(m_n - m_p - m_e)c^2 + a_C A^{-1/3} + 4a_{sym}}{2a_C A^{-1/3} + 8a_{sym} A^{-1}}. \tag{4.13}$$

The most stable nuclide is given by the integer value closest to Z_{min}.

The physical basis for the instability of nuclides with Z on either side of the minimum can be described as follows: Nuclei with Z less than Z_{min} have too many neutrons (for the number of protons) and the instability can be explained by the corresponding large value of the symmetry term in equation (4.12). Nuclei with Z greater than Z_{min} have too many protons (for the number of neutrons) and the instability can be explained by a combination of the symmetry and Coulomb terms in equation (4.12).

In the first case, a negative β-decay (β^- decay) process occurs:

$$^A_Z X^N \rightarrow ^A_{Z+1} Y^{N-1} + e^- + \bar{\nu}_e \tag{4.14}$$

where an electron is emitted during the process. Here a nucleus of element X has been converted to a nucleus of element Y and we have written the corresponding values of Z as subscripts before the element name and the values of N as superscripts after the element name in order to help keep track of all the nucleons in the problem. This process is the result of the weak interaction (since neutrinos are involved—see Chapter 2) and an inspection of equation (4.14) will show that charge, lepton number, and baryon number are conserved. In fact, the appearance of the antineutrino on the right-hand side of the equation is necessary to conserve lepton number. The basic process described by equation (4.14) is the conversion of a neutron to a proton:

$$n \rightarrow p + e^- + \bar{\nu}_e. \tag{4.15}$$

The second case corresponds to positive β-decay (β^+ decay) and is

$$^A_Z X^N \rightarrow ^A_{Z-1} Y^{N+1} + e^+ + \nu_e \tag{4.16}$$

where a positron (or antielectron) is emitted. Again the neutrino is necessary to satisfy conservation of lepton number. This decay corresponds to the conversion of a proton to a neutron or

$$p \rightarrow n + e^+ + \nu_e. \tag{4.17}$$

Table 4.1 | Properties of Nuclides with $A = 135$ (ec = electron capture)

Element	N	Z	Mass (u)	Lifetime	Decay Mode	Daughter	Decay Energy (MeV)
I	82	53	134.909823	9.6 h	β^-	^{135}Xe	2.51
Xe	81	54	134.907130	13.1 h	β^-	^{135}Cs	1.16
Cs	80	55	134.905885	4.3×10^6 y	β^-	^{135}Ba	0.21
Ba	79	56	134.905665	stable	—	—	—
La	78	57	134.906953	28.0 h	ec	^{135}Ba	1.20
Ce	77	58	134.909117	25.3 h	ec, β^+	^{135}La	2.02

These β-decay processes, as well as electron capture, which is equivalent to β^+ decay, will be considered in detail in Chapter 9. Table 4.1 gives some of the relevant properties for the nuclides shown in Figure 4.3. A general observation can be made concerning the lifetimes of the nuclides described in the table. As the nuclides decay towards ^{135}Ba from either side the lifetimes become longer. This is directly related to the decreasing difference in mass between the parent and daughter nucleus. This same feature is seen in subsequent decays described in this chapter.

On the basis of Figure 4.3 it is expected that for a given value of odd A there should be uniquely one β stable nuclide. Since odd A can result from N and Z odd–even or even–odd we would expect approximately equal numbers of these two types of nuclei to exist. This is in agreement with Table 3.2.

The situation for even A, which occurs for odd–odd or even–even nuclei, is much more complex because the pairing term is nonzero. In general we expect two parabolas for m as a function of Z, one shifted up (the odd–odd parabola) and one shifted down (the even–even parabola) as illustrated in Figure 4.4 for $A = 140$. The stable nucleus again occurs for the minimum value of mass and the figure shows that as nuclei with too few or too many protons decay by β-decay processes they alternate from the odd–odd parabola to the even–even parabola. For the case shown ^{140}Ce with $Z = 58$ lies very close to the minimum in m and represents the β stable nuclide with $A = 140$. Properties of the decay process shown in the figure are given in Table 4.2.

Another situation for an even A nucleus ($A = 128$) is illustrated in Figure 4.5. Here there are two stable nuclei on the even–even parabola; ^{128}Te and ^{128}Xe. ^{128}I can decay by either β^- decay to ^{128}Xe or by β^+ decay to ^{128}Te. This means that it becomes more stable either by converting a neutron to a proton or a proton to a neutron. This can be understood on the basis of the semiempirical mass formula because either process will change an odd–odd nucleus to an even–even nucleus.

Figure 4.4 | Mass Parabola for $A = 140$ Showing One Stable Nuclide with $Z = 58$

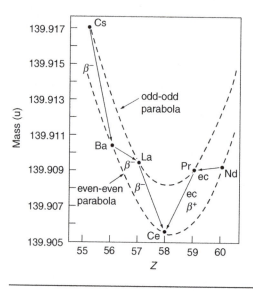

However, the real implications of this pairing behavior will become obvious in the next chapter. Relevant nuclear properties are given in Table 4.3.

A third possibility for an even A nucleus is shown in Figure 4.6 ($A = 130$). Here there are three stable nuclei on the even–even parabola; ^{130}Te, ^{130}Xe, and ^{130}Ba. In this case ^{130}Cs can decay by either β^- or β^+ decay to ^{130}Xe or

Table 4.2 | Properties of Nuclides with $A = 140$ (ec = electron capture)

Element	N	Z	Mass (u)	Lifetime	Decay Mode	Daughter	Decay Energy (MeV)
Cs	85	55	139.917338	95 s	β^-	^{140}Ba	6.29
Ba	84	56	139.910518	18.3 d	β^-	^{140}La	1.03
La	83	57	139.909471	58.0 h	β^-	^{140}Ce	3.76
Ce	82	58	139.905433	stable	—	—	—
Pr	81	59	139.909071	4.9 m	ec, β^+	^{140}Ce	3.39
Nd	80	60	139.909036	4.75 d	ec	^{140}Pr	0.22

Figure 4.5 | Mass Parabola for $A = 128$ Showing Two Stable Nuclides with $Z = 52$ and 54

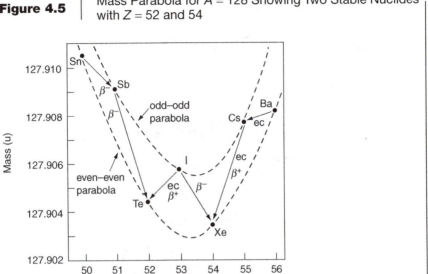

^{130}Ba, respectively. ^{130}I decays by β^- decay to ^{130}Xe. Although it is energetically favorable for ^{130}I to decay to ^{130}Te, this process has not been observed. Relevant nuclear properties are given in Table 4.4. The stability of nuclides such as ^{130}Te, which are at a local minimum on the mass curve will be discussed further in Chapter 9 in the context of double β-decay.

Table 4.3 | Properties of Nuclides with $A = 128$ (ec = electron capture)

Element	N	Z	Mass (u)	Lifetime	Decay Mode	Daughter	Decay Energy (MeV)
Sn	78	50	127.910467	85 m	β^-	^{128}Xe	1.30
Sb	77	51	127.909072	15.6 m	β^-	^{128}Cs	4.29
Te	76	52	127.904463	stable	—	—	—
I	75	53	127.905810	36 m	ec, β^+	^{128}Te	1.26
					β^-	^{128}Xe	2.12
Xe	74	54	127.903531	stable	—	—	—
Cs	73	55	127.907762	5.5 m	ec, β^+	^{128}Xe	3.94
Ba	72	56	127.908237	3.46 d	ec	^{128}Cs	0.44

Figure 4.6 | Mass Parabola for $A = 130$ Showing Three Stable Nuclides with $Z = 52$, 54, and 56

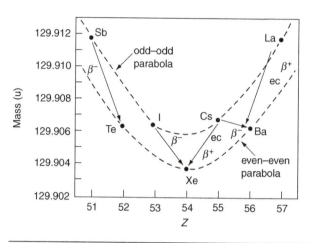

From Figures 4.4 through 4.6, it is seen that either one, two, or three even–even β-stable nuclei can occur for a given (even) value of A. This feature is responsible for the anomalously large number of known stable even–even nuclei as was indicated in Table 3.1.

Table 3.1 indicated that there are four β-stable nuclei that are odd–odd. If we examine a situation as shown in Figure 4.5 where a nuclide is situated near the minimum in the odd–odd parabola (^{128}I) it is difficult to image how any

Table 4.4 | Properties of Nuclides with $A = 130$ (ec = electron capture)

Element	N	Z	Mass (u)	Lifetime	Decay Mode	Daughter	Decay Energy (MeV)
Sb	79	51	129.911546	10 m	β^-	^{130}Te	4.95
Te	78	52	129.906229	stable	—	—	—
I	77	53	129.906713	17.6 h	β^-	^{130}Xe	2.98
Xe	76	54	129.903509	stable	—	—	—
Cs	75	55	129.906753	43.2 m	ec, β^+ β^-	^{130}Xe ^{130}Ba	3.02 0.51
Ba	74	56	129.906282	stable	—	—	—
La	73	57	129.912320	12.5 m	ec, β^+	^{130}Ba	5.62

Figure 4.7 | Mass Parabola for $A = 14$ Showing the Existence of a Stable Odd–Odd Nuclide

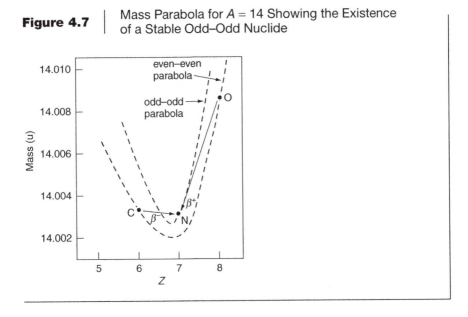

odd–odd nuclei could be stable, since the odd–odd nucleus can decay to the even–even nuclei on either side. If we examine the shape of the mass parabola as given by equation (4.12), we learn that the parabola becomes narrower as A decreases. Thus for small A we can have the situation as shown in Figure 4.7 for $A = 14$. Here the sides of the parabolas are sufficiently steep that the minimum in the odd–odd parabola lies below the adjacent points on the even–even parabola and the odd–odd nucleus ^{14}N is the stable $A = 14$ nuclide. Other situations can exist, for example ^2H, where β-decay cannot occur because the daughter nucleus does not form a bound state.

The predictions of the semiempirical mass formula can be viewed in the context of the distribution of stable nuclides as was shown in Figure 3.1. In this figure, each particular point in N-Z space corresponds to a particular value of A. Constant A isobars are represented in the figure by lines parallel to the $A = 100$ line shown. Thus in a three-dimensional plot with Figure 3.1 as the xy plane and mass plotted on the z-axis, mass parabolas for constant A lines would form a parabolic surface with the minimum following the stability line in the N-Z plane. This is referred to as the β-stability valley.

4.4 NUCLEON SEPARATION ENERGIES

A measure of nuclear stability that will be discussed in some detail in the next chapter, is the energy required to remove one nucleon from the nucleus. This is not the same for neutrons and protons. We can define two quantities, S_n and S_p, the neutron separation energy and the proton separation energy, respectively.

The removal of a neutron from a nucleus corresponds to the process

$$
{}_Z^A X^N \rightarrow {}_Z^{A-1} X^{N-1} + n
\tag{4.18}
$$

where the number of protons (and hence the identity of the element) has not changed but the number of neutrons (and hence the value of A) has decreased by one. The energy required to produce this process (assuming that it is endothermic and does not occur spontaneously) is given in terms of the masses of the components in equation (4.18) as

$$
S_n = \left[m\left({}_Z^{A-1} X^{N-1} \right) + m_n - m\left({}_Z^A X^N \right) \right] c^2.
\tag{4.19}
$$

Here there is no change in the number of atomic electrons, therefore atomic masses (rather than nuclear masses) can be used because the electron masses will cancel out.

The removal of a proton from the nucleus corresponds to the process

$$
{}_Z^A X^N \rightarrow {}_{Z-1}^{A-1} Y^N + p
\tag{4.20}
$$

and the separation energy is given by

$$
S_p = \left[m\left({}_{Z-1}^{A-1} Y^N \right) + m_p + m_e - m\left({}_Z^A X^N \right) \right] c^2.
\tag{4.21}
$$

where the electron mass is included to allow for the use of atomic masses.

Table 4.5 gives the nucleon separation energies for some light nuclei. It is clear that S_n and S_p can, in some cases, be substantially different. On the basis of the shell model discussed in the next chapter, this behavior is readily expected. These nucleon separation processes can be related to the predictions of the semiempirical mass formula. Figure 4.8 shows a typical example for neutron separation from a moderately heavy nucleus. In this example the separation of a neutron from an odd A nucleus is considered, however, an analogous diagram can be constructed for the even A case. No specific Z values are indicated but the horizontal axis does illustrate that Z alternates between even (e) and odd (o) values. Mass parabolas for A and $A - 1$ are plotted on the same graph. The mass parabolas for $A - 1$ are shifted upward from their actual values by the mass of one neutron in order to properly account for the energetics of equation (4.19). On the average (that is, pairing term ignored) the mass of the most stable nuclide with $A - 1$ nucleons plus the mass of one nucleon is about 8 MeV greater than the mass of the most stable nuclide with A nucleons. This is just the mass associated with the average binding energy of one nucleon. In general this means that neutron separation is an endothermic process and does not occur spontaneously. Neutron separation results in a transition from the odd A parabola to one of the even A parabolas. Since there is no change in Z the transition is represented by a vertical line as shown in the figure. If the initial nucleus was an odd–even nucleus then the final nucleus is an even–even nucleus; if the initial nucleus is even–odd then the final one is odd–odd. This is shown in the figure

Table 4.5 | Binding Energies and Nucleon Separation Energies for Some Light Nuclei

Nuclide	N	Z	B/A (MeV)	S_n (MeV)	S_p (MeV)
^{2}H	1	1	1.11	2.22	—
^{3}He	1	2	2.57	—	5.49
^{4}He	2	2	7.08	20.58	19.81
^{6}Li	3	3	5.33	5.66	4.59
^{7}Li	4	3	5.61	7.25	9.98
^{9}Be	5	4	6.46	1.67	16.89
^{10}B	5	5	6.48	8.44	6.59
^{11}B	6	5	6.93	11.45	11.22
^{12}C	6	6	7.68	18.72	15.96
^{13}C	7	6	7.47	4.95	17.53
^{14}N	7	7	7.48	10.55	7.55
^{15}N	8	7	7.70	10.83	10.21
^{16}O	8	8	7.98	15.66	12.13
^{17}O	9	8	7.75	4.14	13.78
^{18}O	10	8	7.77	8.04	15.94

Figure 4.8 | Representation of Neutron Separation Energies

Arrows show the transitions from the odd A parabola to the even $A − 1$ parabolas for odd–even and even–odd starting nuclei.

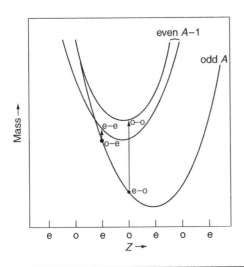

Figure 4.9 | Representation of Proton Separation Energies

Arrows show the transitions from the odd A parabola to the even $A - 1$ parabolas for even–odd and odd–even starting nuclei.

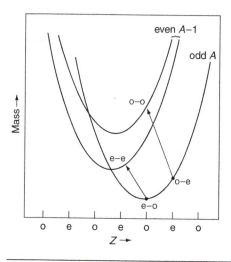

for both initial even and odd values of Z. It is clear that the energy involved in these two cases is substantially different and this is consistent with the indications in Table 4.5 and the subsequent discussion in Chapter 5.

For proton separation the situation is somewhat more complex as there is a change in the value of Z. This is illustrated in Figure 4.9 for transitions from odd–even to odd–odd and from even–odd to even–even nuclei. It is clear from the figures that neutron separation energies can, in some cases, be much smaller than proton separation energies and this is consistent with the data in Table 4.5.

In a few cases very neutron-rich nuclei can decay spontaneously from their excited states by neutron emission. This is most commonly observed in fission byproducts (see Chapter 12). Neutron emission is also known from excited states (see Chapters 6 and 11) in nuclei with small A where the mass parabolas are very steep.

Problems

4.1. Using the semiempirical mass formula and the values of the parameters given in the text, calculate the value of Z for the minimum in the mass parabola for $A = 91$ and $A = 123$. Discuss these results in the context of the known stability for nuclei with these values of A.

4.2. (a) Use the semiempirical mass formula to calculate the binding energy per nucleon for ^{3}H, ^{4}He, ^{64}Cu, and ^{119}Sn.

(b) Use measured atomic masses to calculate the binding energy per nucleon for the three nuclides in part (a).

(c) Discuss the results of parts (a) and (b) and comment on the validity of the liquid drop model.

4.3. Derive an expression for the internal (Coulomb) energy of a uniformly charged sphere of radius r and total charge $+Ze$. Compare this with the form of the Coulomb term in the semiempirical mass formula.

4.4. (a) Using measured atomic masses calculate the binding energies of ^{13}C and ^{13}N. These nuclei are referred to as mirror nuclei.

(b) On the basis of the liquid drop model, describe the reasons for the differences observed in part (a).

(c) From these results determine the radius of a ^{13}C nucleus.

4.5. Using the values of the mass defects for ^{4}He, ^{18}F, and ^{198}Pt, of +2.42 MeV, +0.872 MeV and −29.91 MeV, respectively, calculate the atomic masses of these nuclides. Check your answers by comparison with tabulated atomic mass values.

4.6. Using measured atomic masses calculate the neutron and proton separation energies for ^{235}U, ^{236}U, ^{235}Np, and ^{236}Np. Comment on any trends that you observe in these results.

4.7. Use the semiempirical mass formula to estimate the relative importance of the volume, surface, Coulomb, and symmetry terms for nuclei with $A = 3, 19, 99$, and 201.

4.8. Use the semiempirical mass formula to estimate N/Z for stable nuclei with $A = 3, 19, 99$, and 201.

4.9. Identify all stable odd–odd nuclides. Sketch and discuss mass diagrams (see Figure 4.7) for each of these cases.

4.10. (a) Rewrite the semiemprical mass formula in terms of A and N.

(b) Use the results of part (a) to derive an expression for the two-proton separation energy (that is, the energy required to remove two protons from the nucleus).

(c) Why is it more convenient to deal with the two-proton separation energy in this case than the one-proton separation energy?

4.11. (a) Plot the atomic masses of all nuclides given in Appendix B with $A = 64$.

(b) Use this figure to estimate the value of a_p in the semiempirical mass formula and compare it with the accepted value given in the text.

5 CHAPTER | The Shell Model

5.1 OVERVIEW OF ATOMIC STRUCTURE

Previous chapters stressed the importance of neutron and proton pairing. This is analogous in some ways to the atomic situation as we are dealing with a system of bound fermions. However, there are important differences and these are summarized in Table 5.1. In the atomic case we need only consider the electromagnetic force, which is well known, but in the nuclear case the strong interaction must also be included and is expected to be the dominant contribution to the nuclear potential. In order to understand the similarities and differences between the behavior of atoms and nuclei we begin with a brief look at some of the characteristics of electrons in an atom.

Consider a single electron in the coulombic potential of the positive charge of the nucleus, $+eZ$. This system is referred to as a hydrogen-like atom and the energy levels are illustrated in Figure 5.1(a). These energy levels are characterized by a single quantum number n. The actual situation of a many-electron atom is substantially more complex. Principally, the electrostatic interactions affecting an electron are due not only to the charge of $+eZ$ on the nucleus, but also the charges of the other $(Z - 1)$ atomic electrons. This gives rise to the splitting of the energy levels as illustrated in Figure 5.1(b). Additional fine structure splitting of the energy levels (not shown in the figure) results from the magnetic coupling between the spins and orbital angular momenta of the electrons. Mathematically the problem of calculating the energy levels in a multi-electron atom is complex and methods are described in most quantum

Table 5.1	Similarities and Differences in the Modeling of Atomic Electrons and Nucleons	

Property	Electrons	Nucleons
Type of particles	fermions	fermions
Identity of particles	all electrons	neutrons and protons
Charges	all charged	some charged
Occupancy considerations	Pauli principle	Pauli principle
Interactions	electromagnetic	strong and electromagnetic

mechanics texts. Each energy level is defined in terms of the principal quantum number n and the quantum number associated with the orbital angular momentum of the level, l. According to conventional spectroscopic notation the letters s, p, d, f, g, h... correspond to the values of $l = 0, 1, 2, 3, 4, 5...$. The degeneracy of each energy level is given by the product of the spin degeneracy, 2, and the orbital angular momentum degeneracy, $2l + 1$. These result from

Figure 5.1	Atomic Energy Levels for $n = 1$ to 4 in (a) a Hydrogen-Like Atom and (b) a Many-Electron Atom	

Note that the energy axis is not to scale.

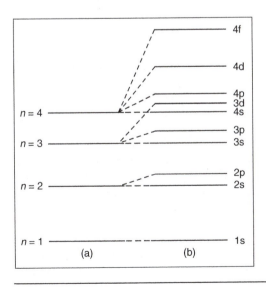

Table 5.2 | Occupancy of Atomic Energy Levels in Order of Increasing Energy

Shell	n	l	Notation	Degeneracy $2(2l+1)$	Accumulated Occupancy
K	1	0	1s	2	2
L	2	0	2s	2	4
L	2	1	2p	6	10
M	3	0	3s	2	12
M	3	1	3p	6	18
N	4	0	4s	2	20
N	3	2	3d	10	30
N	4	1	4p	6	36
O	5	0	5s	2	38
O	4	2	4d	10	48
O	5	1	5p	6	54
P	6	0	6s	2	56
P	4	3	4f	14	70
P	5	2	5d	10	80
P	6	1	6p	6	86

the allowed values of the z-component of the spin, $m_s = -1/2, +1/2$ and the z-component of the orbital angular momentum, $m_l = -l - l + 1 \ldots l - 1, l$. The total degeneracy, $2(2l + 1)$ gives the electron occupancy of each level as indicated in Table 5.2.

As Figure 5.1 illustrates there are clusters of energy levels with gaps between them. These clusters of energy levels are referred to as shells and are given the designations K-shell, L-shell, M-shell, etc., with increasing energy. Atoms with filled electronic shells are particularly stable and this is manifested as an anomalously large ionization energy. On the other hand, atoms with one electron more than necessary to fill a shell have very low ionization energies as this final electron is very weakly bound. Figure 5.2 shows the measured ionization energy as a function of atomic number, Z. Peaks in the ionization energy clearly correspond to filled electron shells as indicated in the table. The number of electrons corresponding to these particular configurations are referred to as magic numbers, in this case the atomic magic numbers. Table 5.2 and Figure 5.2 show that the atomic magic numbers are 2, 10, 18, 36, 54, 86.... Although the problem

Figure 5.2 | Electron Ionization Energies as a Function of Z

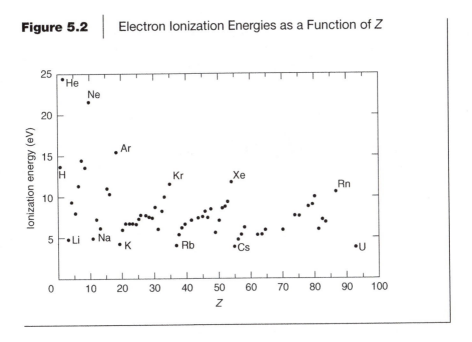

of calculating the energy levels of multielectron atoms is mathematically complex, the theoretically predicted atomic magic numbers are in agreement with the experimental observations.

5.2 EVIDENCE FOR NUCLEAR SHELL STRUCTURE

There are indications of a shell structure, and corresponding magic numbers, associated with the nuclear energy levels that are analogous to the electronic case. The most straightforward evidence for this kind of behavior might be a consideration of nucleon separation energy as a function of the number of nucleons. This kind of approach for nuclei is somewhat complicated as there are two kinds of nucleons and, as shown in the last chapter, their separation energies can be quite different. Thus it is not apparent whether A, N, or Z or some combination of these would be most relevant in determining nuclear stability. Some experimental observations are discussed below.

Binding Energy   Since the derivation of the semiempirical mass formula did not include any information about the shell structure of the nucleus, it is expected that deviations of experimental data from these model predictions can be indicative of shell effects. Figure 5.3 shows the difference between the liquid drop model predictions and actual measurements of B/A as a function of A. Clear deviations from the model prediction can be seen for certain values of A.

Figure 5.3 | Differences Between the Measured Binding Energy per Nucleon and the Value Predicted by the Liquid Drop Model as a Function of *A*

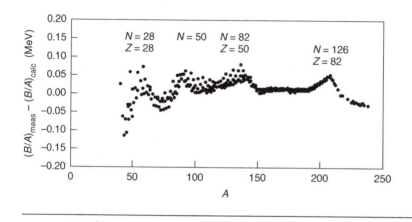

Since stable nuclei have a reasonably well-defined relationship between A, N, and Z these anomalies correspond to specific N and/or Z values as indicated in the figure. There is evidence from the figure that there is excess binding energy for nuclei with N or $Z = 28$, 50, 82, and 126 indicating particular stability for these nuclei. For light nuclei the behavior is illustrated in Figure 5.4. The binding energy per nucleon is shown for nuclei with $N = Z$. The general increase in B/A

Figure 5.4 | Binding Energy per Nucleon for Light Nuclei with $N = Z$

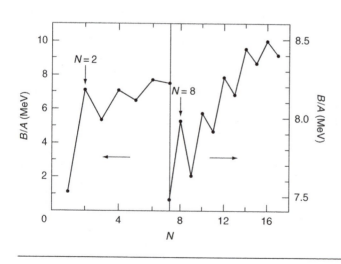

Figure 5.5 | Change in the Measured Nuclear Radius for a Change in Neutron Number $\Delta N = 2$ Normalized to the Change Predicted by Equation (3.23)

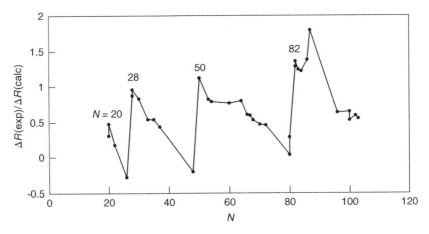

Data from E. B. Shera *et al.*, *Phys. Rev. C14* (1976), 731. Copyright 1976 by the American Physical Society.

as predicted by the semiempirical mass formula is seen, but slightly higher values are indicated for $N = 2$ and 8. The pairing effects are also obvious.

Nuclear Radius   Deviations in the nuclear radius from the simple behavior predicted by equation (3.23) are indicative of the nuclear shell structure. Figure 5.5 illustrates this behavior and shows the existence of magic numbers for 20, 28, 50, and 82 neutrons.

Number of Stable Nuclides   Although isotopes of an element all have the same value of Z, several values of N are possible. Thus several stable nuclides with the same N (or isotones) can exist. A larger number of stable nuclides are possible for certain values of N as illustrated in Figure 5.6. This indicates that nuclei with $N = 20$, 28, 50, and 82 have particular stability.

Neutron Absorption Cross Section   The cross section for fast neutron absorption, σ, as a function of N is shown in Figure 5.7. If a specific nucleon configuration is particularly stable we expect that there will be a low probability of absorbing an additional neutron, and hence a low cross section. The figure shows that anomalously low cross sections occur for $N = 50$, 82, and 126.

First Excited State Energies   Nuclei possess excited states and these can be populated during various decay processes or reactions. Excited nuclear states

Figure 5.6 | Number of Stable Isotones as a Function of N (Even)

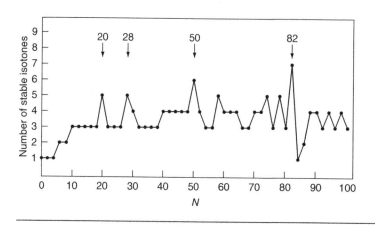

Figure 5.7 | Absorption Cross Sections for 1 MeV Neutrons

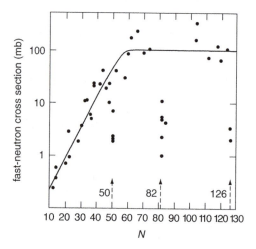

Data from D. J. Hughes and D. Sherman, *Phys. Rev.* 78 (1950), 632. Copyright 1950 by the American Physical Society.

can be viewed much in the same way as excited atomic electron states, by the occupation of a normally unoccupied higher energy level by one of the nucleons. Details of excited states will be discussed in Chapter 6. If a particular nuclear configuration is particularly stable then it is expected that a larger amount of

Figure 5.8 | Energies of the First Excited 2⁺ States of Even–Even Nuclei as a Function of A and Z

The relevance of the first excited 2⁺ state is discussed in Chapter 6.

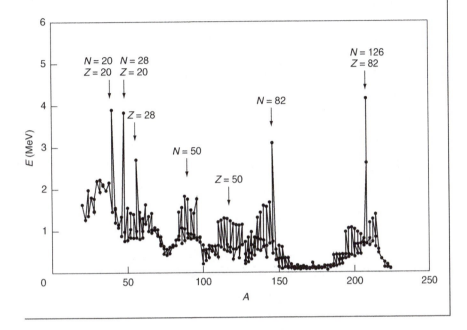

energy would be required to introduce a transition to an excited state. This is the case for nuclei with filled shells and is shown in Figure 5.8. Clear peaks are seen for N or Z = 20, 28, 50, 82, and 126. The nuclei with Z = 20 (Ca) are of particular interest. Figure 5.9 shows the excited state energies for even–even Ca. Clear peaks are seen for N = 20 and N = 28. In fact these two nuclei have both N and Z magic and are referred to as doubly magic.

Other nuclear properties that show characteristic behavior as a function of nucleon number will be discussed in subsequent chapters. Overall, observations of properties related to nuclear stability can be summarized in terms of the so-called nuclear magic numbers, 2, 8, 20, 28, 50, 82, and 126, and indicate that nuclei with N or Z equal to a magic number show particular stability. It would be desirable to develop a model that, along the lines of the atomic situation described above, would correctly predict these nuclear magic numbers. The use of an appropriate potential in the Schrödinger equation that could predict these magic numbers would provide evidence that our understanding of the nature of the strong interaction is valid. The remainder of this chapter is devoted to the development of such a model.

Figure 5.9 | Energies of the First Excited 2⁺ States of Even–Even Isotopes of Ca ($Z = 20$) as a Function of N

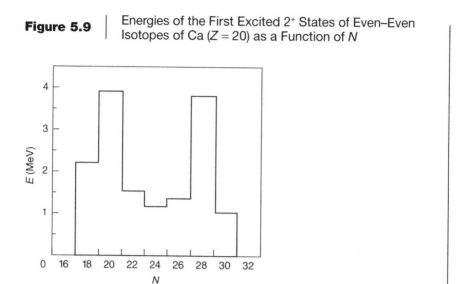

5.3 THE INFINITE SQUARE WELL POTENTIAL

Unlike the atomic case we do not have a simple analytical form of the potential for the nuclear interaction. However, some simple models can be surprisingly effective for describing the behavior of nuclei. We will begin with a consideration of the energy levels for neutrons, which is slightly simpler, and will discuss the situation for protons (which are subject to coulombic interactions) in somewhat less detail later. To some extent the solution to this problem is a matter of refinement. The procedure is as follows:

1. Approximate the mutual interaction between nucleons by a single particle potential, called the nuclear potential.
2. Solve the Schrödinger equation for the energy eigenstates.
3. Use these results to determine the corresponding magic numbers and compare these with the experimental numbers given above.
4. If the calculated results do not agree with the experiment, revise the form of the potential and try again.

The solution obtained in this manner is sometimes referred to as the single-particle model solution for the nucleus as it considers the behavior of each nucleon in a fixed nuclear potential. We begin with the simplest possible potential that could give reasonable results: the infinite spherical square well. This potential will account for the fact that the nucleons are well bound within the nucleus, that the nucleon density is relatively constant within the nucleus, and that this density is more or less independent of the total nuclear mass.

The time independent Schrödinger equation in three dimensions is expressed as

$$-\frac{\hbar^2}{2m}\nabla^2\psi + V\psi = E\psi. \tag{5.1}$$

For a spherically symmetric potential we assume that the solution is separable and can be written as

$$\psi(r,\theta,\phi) = R(r)\Theta(\theta)\Phi(\phi). \tag{5.2}$$

The radially dependent potential only affects the solution for $R(r)$ meaning that the form of the Θ and Φ functions can be determined independently of $V(r)$. Substituting (5.2) into (5.1) gives the differential equation in Φ as

$$\frac{d^2\Phi}{d\phi^2} + m_l^2\Phi = 0 \tag{5.3}$$

where m_l is an integer that takes on values $m_l = 0, \pm1, \pm2, \dots$. The solutions to this equation are of the form

$$\Phi_{m_l}(\phi) = \frac{1}{\sqrt{2\pi}}e^{im_l\phi}. \tag{5.4}$$

The Schrödinger equation for Θ is of the form

$$\frac{1}{\sin\theta}\frac{d}{d\theta}\left[\sin\theta\frac{d\Theta}{d\theta}\right] + \left[l(l+1) - \frac{m_l^2}{\sin^2\theta}\right]\Theta = 0. \tag{5.5}$$

where $l = 0, 1, 2, \dots$ and m_l must take on integer values, $0, \pm1, \pm2, \dots \pm l$. The solution to (5.5) is a polynomial in $\sin\theta$ or $\cos\theta$ (an associated Legendre polynomial) and can be combined with equation (5.4) to give the angular dependence in terms of the spherical harmonics

$$Y_{lm_l}(\theta,\phi) = \Theta_{lm_l}(\theta)\Phi_{m_l}(\phi). \tag{5.6}$$

Typical solutions are given as a function of l and m_l in Table 5.3.

The energy eigenvalues are obtained from the radial part of the Schrödinger equation:

$$-\frac{\hbar^2}{2m}\left[\frac{d^2R}{dr^2} + \frac{2}{r}\frac{dR}{dr}\right] + \left[V(r) + \frac{l(l+1)\hbar^2}{2mr^2}\right]R = ER \tag{5.7}$$

where $l = 0, 1, 2, 3, \dots$.

An infinite spherical square well of radius a is defined as

$$V(r) = 0 \qquad r < R_0$$
$$V(r) = \infty \qquad r \geq R_0. \tag{5.8}$$

Table 5.3 | Example of Spherical Harmonics for Small Values of l

l	m_l	$Y_{lm_l}(\theta, \phi)$
0	0	$(1/4\pi)^{1/2}$
1	0	$(3/4\pi)^{1/2} \cos\theta$
1	± 1	$\mp (3/8\pi)^{1/2} \sin\theta \, e^{\pm i\phi}$
2	0	$(5/16\pi)^{1/2} (3\cos^2\theta - 1)$
2	± 1	$\mp (15/8\pi)^{1/2} \sin\theta \cos\theta \, e^{\pm i\phi}$
2	± 2	$(15/32\pi)^{1/2} \sin^2\theta \, e^{\pm 2i\phi}$

Table 5.4 | Some Examples of Spherical Bessel Functions

l	$j_l(kr)$
0	$\sin kr/(kr)$
1	$[\sin kr/(kr)^2] - [\cos kr/(kr)]$
2	$[3\sin kr/(kr)^3] - [3\cos kr/(kr)^2] - [\sin kr/(kr)]$

Figure 5.10 | Spherical Bessel Functions for $l = 0, 1,$ and 2

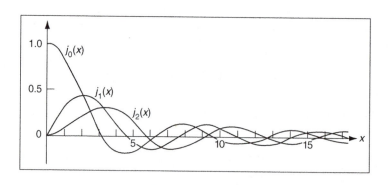

Solutions to equation (5.7) with $V = 0$ are the spherical Bessel functions, $j_l(kr)$, as given in Table 5.4 where $k = \sqrt{2mE/\hbar^2}$. The energy eigenvalues are found from these solutions by the application of the boundary condition $j_l(kR_0) = 0$. Examples of spherical Bessel functions are shown in Figure 5.10. The boundary

Table 5.5	The First Four Zeros of the Spherical Bessel Functions as a Function of l

l	$z_n(l)$
0	3.1416, 6.2832, 9.4248, 12.5664, ...
1	4.4934, 7.7253, 10.9041, 14.0662, ...
2	5.7635, 9.0950, 12.3229, 15.5146, ...
3	6.9879, 10.4171, 13.6980, 16.9236, ...
4	8.1826, 11.7049, 15.0397, 18.3013, ...
5	9.3558, 12.9665, 16.3547, 19.6532, ...
6	10.5128, 14.2074, 17.6480, 20.9835, ...
7	11.6570, 15.4313, 18.9230, 22.2953, ...

condition is satisfied for specific values of kR_0, that is, the zeros of the spherical Bessel function, $z_n(l)$. From the definition of k it is easily seen that the energy eigenvalues are given in terms of the $z_n(l)$ as

$$E = \frac{\hbar^2 z_n(l)^2}{2R_0^2 m} = E_0 \frac{z_n(l)^2}{\pi^2} \tag{5.9}$$

where E_0 is defined as $E_0 = \hbar^2\pi^2/2R_0^2 m$. Values of the $z_n(l)$ can be tabulated using the expressions given in Table 5.4 and some of these are summarized in Table 5.5. For a given value of l there is a series of zeros as given in the table and these can be seen as the zero crossing points of the spherical Bessel functions in Figure 5.10. These different $z_n(l)$ correspond to different values of the principal quantum number, n. Tabulating calculated values of E in increasing order from each of the $z_n(l)$, allows the energy levels to be obtained as given in Table 5.6 as a function of n and l. The spectroscopic notation as shown in the table has been adapted from the atomic nomenclature discussed before where the l values 0, 1, 2, 3, 4,... are designated s, p, d, f, g,... . Analogous to the atomic case each of these states is degenerate in m_l where the degeneracy is given by $(2l + 1)$ and also in spin where the degeneracy is 2 ($m_s = \pm^1/_2$). The resulting neutron-level occupancy and accumulative number of neutrons is given in the table and the energy levels are illustrated in Figure 5.11. An inspection of the energy of the various levels shows that it is a little difficult to see clearly defined shells as could be observed for atomic electrons. However, it is clear that aside from 2, 8, and 20, the accumulated occupancies given in the table cannot correspond to the observed nuclear magic numbers.

Table 5.6 | Energies and Occupancy of Nucleon States for the Infinite Square Well Potential

n	l	Notation	E/E_0	Occupancy	Accumulated Occupancy
1	0	1s	1.00	2	2
1	1	1p	2.05	6	8
1	2	1d	3.37	10	18
2	0	2s	4.00	2	20
1	3	1f	4.96	14	34
2	1	2p	6.04	6	40
1	4	1g	6.78	18	58
2	2	2d	8.38	10	68
1	5	1h	8.88	22	90
3	0	3s	9.00	2	92
2	3	2f	10.99	14	106
1	6	1i	11.20	26	132
3	1	3p	12.05	6	138
1	7	1j	13.76	30	168

5.4 OTHER FORMS OF THE NUCLEAR POTENTIAL

Because the nuclear model as described above does not yield the correct magic numbers, it might seem reasonable to modify the nuclear potential. Because we do not expect infinite potential walls at the edges of the nucleus, it would be reasonable to consider a finite square well. The implementation of this model follows along the lines of the infinite well, and as long as the potential remains spherically symmetric, the energy eigenvalues are determined from a consideration of the radial equation. In this case the nucleon wave functions will penetrate the side of the well and the boundary conditions for the wave functions must be satisfied at $r = R_0$. In general, this has the effect of uniformly lowering the energy levels from those calculated for the infinite square well, but does not help to predict the correct magic numbers as illustrated in Figure 5.11.

From a physical standpoint the most reasonable nuclear potential would be one that looked like the total nucleon density in the nucleus as shown in Figure 3.9. This can be accomplished by using a potential in the Schrödinger equation that is proportional to the Woods-Saxon function given by equation (3.18). In practice, other analytical forms for the potential are mathematically

Figure 5.11 | Nuclear Energy Levels for an Infinite Square Well, a Finite Square Well, a Square Well with Rounded Edges, and a Square Well with Rounded Edges Including Spin Orbit Coupling

The numbers on the right hand side of the figure show the cumulative occupancy for the final case. The harmonic oscillator energy levels are shown for comparison.

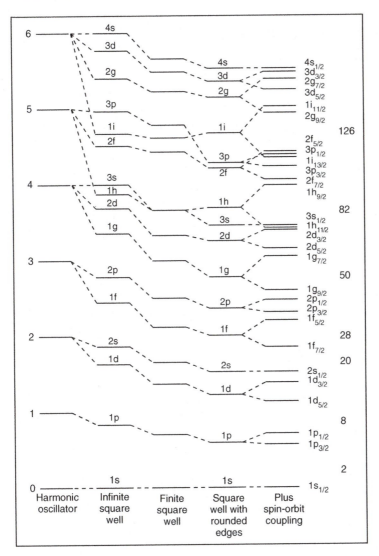

From B. T. Feld, *Ann. Rev. Nucl. Sci.* 2 (1953), 239. Reprinted with permission from the *Annual Review of Nuclear Science*, Volume 2 © 1953 by Annual Reviews, www.annualreviews.org.

easier to deal with. One that is often used is a finite square well with exponential sides. The details of the shape of the potential have relatively little influence on the success of this model. The energy levels for a nuclear potential described by a rounded square well are compared with those found for the infinite square well in Figure 5.11. Some minor changes in the ordering and spacing of the energy levels occur, but the model still fails to predict the correct magic numbers.

5.5 SPIN-ORBIT COUPLING

It is clear that the details of the potential used in the Schrödinger equation have only a minor influence on the ordering of the nuclear energy levels and that any reasonable potential will probably be unable to explain the experimentally observed magic numbers. The proper description of experimental data, therefore, requires the introduction of an additional factor. It is well known that the correct description of atomic energy levels requires a consideration of the interaction between the electron's spin and its orbital angular momentum; the so-called spin-orbit interaction. In a classical sense this interaction can be thought of as a magnetic interaction between the electron's magnetic dipole moment and atom's Coulomb field. Although the spin-orbit interaction is crucial to the proper description of atomic energy levels, its possible introduction into the nuclear problem requires a consideration of the energy scales involved. The typical spacing of electronic energy levels is of the order of a few eV, the typical spacing of nuclear energy levels is closer to an MeV. The spin-orbit interaction as applied to atomic energy levels would be insufficient to have any appreciable effect on nuclear energy levels. However, as scattering experiments suggest, because of the importance of this interaction in nuclei it is reasonable to hypothesize a strong nuclear spin-orbit coupling. Success of the model described below on the basis of this hypothesis provides evidence for the existence of such an interaction. Thus the potential in the Schrödinger equation is modified as

$$V(r) \rightarrow V(r) + f(r)\vec{l} \cdot \vec{s} \qquad (5.10)$$

where $f(r)$ is a function of position and $\vec{s}$ and $\vec{l}$ are the nucleon spin and orbital angular momenta. In the interior of the nucleus where the nucleon density is relatively constant, the spin-orbit interaction cancels out. It is in the edge region of the nucleus where there is a nucleon density gradient that the spin-orbit interaction is important. Thus it is often customary to express the function $f(r)$ in the above as

$$f(r) \propto \frac{1}{r}\frac{dV(r)}{dr}. \qquad (5.11)$$

In the case where there is a strong spin-orbit coupling the energy levels are determined by the total angular momentum, $\vec{j}$, as given by

$$\vec{j} = \vec{l} + \vec{s}. \qquad (5.12)$$

For nucleons $s = \frac{1}{2}$ and the strong spin-orbit coupling imposes the condition on the expectation values for $\vec{j}$:

$$j = l + \frac{1}{2} \quad \text{or} \quad j = l - \frac{1}{2} \tag{5.13}$$

except when $l = 0$, in which case only $j = \frac{1}{2}$ is allowed. Thus $\vec{l}$ and $\vec{s}$ add vectorially (either parallel or antiparallel) to give the total $\vec{j}$. Each energy level is then designated by three quantum numbers n, l, and j and is written in spectroscopic notation as nl_j. This means that each l state is split into two j substates with the two allowed values of j (except for the $l = 0$ s-states, which remain unsplit). The magnitude of the splitting is proportional to $2l + 1$ and $f(r)$. This can be easily shown as follows; the expectation values for l^2, s^2, and j^2 are given by

$$\langle l^2 \rangle = l(l + 1)\hbar^2$$

$$\langle s^2 \rangle = s(s + 1)\hbar^2. \tag{5.14}$$

$$\langle j^2 \rangle = j(j + 1)\hbar^2$$

From equation (5.12) the square of j is found using the cosine rule to be

$$j^2 = l^2 + s^2 + 2\vec{l} \cdot \vec{s}. \tag{5.15}$$

This can be readily solved for $\vec{l} \cdot \vec{s}$ as

$$\vec{l} \cdot \vec{s} = \frac{1}{2}(j^2 - l^2 - s^2). \tag{5.16}$$

The expectation value, determined from equation (5.15), is

$$\langle \vec{l} \cdot \vec{s} \rangle = \frac{\hbar^2}{2}(j(j + 1) - l(l + 1) - s(s + 1)). \tag{5.17}$$

The difference in energy between the two j states can be found from equation (5.10) by using $s = \frac{1}{2}$ and substituting the two values of j from equation (5.13) into equation (5.17). This gives a change in energy of

$$\Delta V \propto 2l + 1. \tag{5.18}$$

Experimental evidence has shown that the sign of $f(r)$ is explicitly negative so that the energy of the state with $j = l - \frac{1}{2}$ is higher than the energy of the state with $j = l + \frac{1}{2}$. The splitting of the energy levels is shown in Figure 5.11.

In order to determine the occupancy of each level it is necessary to properly establish the degeneracy. The degeneracy of the unsplit l levels was given by $2l + 1$ for the allowed values of m_l and 2 from the allowed values of m_s, giving a total degeneracy of $2(2l + 1)$. In the spin-orbit split case, l_z and s_z do not commute with the Hamiltonian, meaning that m_l and m_s are no longer "good" quantum numbers. Because j_z does commute with the Hamiltonian, the relevant "good" quantum number is m_j. The degeneracy of the j states is therefore given by $2j + 1$. As an example consider the 1g state ($n = 1$, $l = 4$ state). This has a degeneracy of $2(2l + 1) = 18$. In the presence of spin-orbit coupling this splits into two states with $j = \frac{9}{2}$ and $j = \frac{7}{2}$ (in order of increasing energy). These then have degeneracies of $2j + 1 = 10$ and $2j + 1 = 8$, respectively, giving the total degeneracy of the initial l state. These features are shown in Figure 5.11. As illustrated in this figure the splitting of l states resulting from the spin-orbit interaction causes a reordering of energy levels and introduces well-defined gaps. These gaps define the various energy shells and, as shown in the figure, correspond to occupancies that match the observed nuclear magic numbers.

5.6 NUCLEAR ENERGY LEVELS

The nuclear shell model that includes a strong spin-orbit coupling has been successful in predicting the correct magic numbers. It is, however, interesting to see how this model can predict the actual energies of the nucleons in their energy levels. A diagram, as illustrated in Figure 5.11, shows the energy levels of the neutrons in a specific nucleus. If we are interested in predicting the behavior of a parameter such as separation energy as a function of N or A it is necessary to calculate the energy level diagram for nuclei of different N (or A) and to understand how these energy levels are occupied. In principle it is easy to see how this will be reflected in the calculation, since the nuclear potential depends on the size of the nucleus. Thus, although the sequence of energy levels is more or less as indicated by Figure 5.11, the actual value of the energy corresponding to each of these levels is a function of the number of nucleons. For a given number of nucleons, the energy of the highest occupied level (the Fermi energy) is found by filling the energy levels with the required number of nucleons as determined by the degeneracy of each state. An example is illustrated in Figure 5.12. The Fermi energy, as determined by the energy of the last nucleon, follows the trends that are observed for the neutron separation energy of various nuclei as a function of A.

It is interesting to note that the spacing of the nuclear energy levels is a sensitive function of the size of the nucleus. Figure 5.12 shows that although there are clear anomalies in the Fermi energy for nuclei with filled shells, the overall change in the Fermi energy between very light nuclei and very heavy nuclei is relatively small. Since heavy nuclei have more nucleons and hence more occupied states, the average spacing between the states below the Fermi energy must decrease as nuclei become heavier. Analogous to the terminology used in solid state physics the number of occupied states per unit energy is referred to as the density of states. This increase in the density of states for heavier (or larger) nuclei is apparent in the simple infinite square well model. As equation (5.9) shows, for a given value of $z_n(l)$, the energy decreases with increasing R_0. This will be considered in more detail in the discussion of excited states in the next chapter.

A final point to consider here is the question of the energy levels of the protons. The above discussion is valid for neutrons but it is clear that the presence of coulombic interactions must alter the form of the potential seen by the protons.

Figure 5.12 | Nuclear Energy Levels for β Stable Odd *A* Nuclei as Predicted by the Shell Model

The energy of the last neutron for odd *N* nuclei is shown by a square and the energy of the last proton for odd *Z* nuclei is shown by a circle.

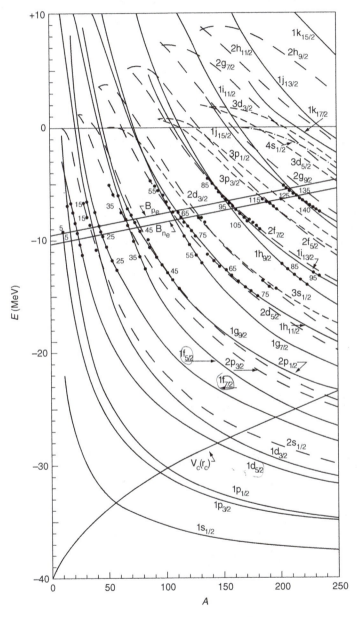

From A. E. S. Green, *Phys. Rev. 104* (1956), 1617. Copyright 1956 by the American Physical Society.

Figure 5.13 | Comparison of Potential Well Shapes for (a) Neutrons and (b) Protons

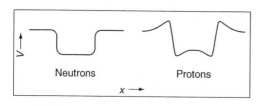

Neutrons   Protons

$x \longrightarrow$

There are basically three effects introduced by the proton charge on the potential seen by these charged particles:

1. A nonzero potential outside the nucleus that follows a $1/r$ dependence as expected outside a spherical distribution of charges
2. A curvature in the bottom of the well as is expected inside a uniform charge distribution
3. A decrease in the depth of the potential well, which accounts for the decreased binding energy for protons as a result of their coulombic repulsion

Figure 5.14 | Comparison of Nuclear Energy Levels for (a) Neutrons and (b) Protons

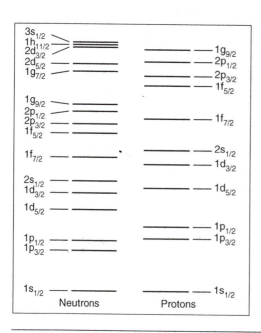

Neutrons   Protons

These features are illustrated in Figure 5.13. The first two characteristics have relatively little effect on the structure of the energy levels, however, this third feature tends to stretch the proton energy levels apart in energy. This is illustrated in Figure 5.14. Although there is not a strict equality, there is a tendency for the neutron Fermi energy and the proton Fermi energy to be approximately the same. It is clear from the figure that if the Fermi energy is small, that is, there are a small number of nucleons, then filling neutron and proton energy levels up to a certain Fermi energy will result in approximately equal numbers of occupied neutron and proton states; hence $N = Z$. For nuclei with a large number of nucleons, filling levels up to the Fermi energy will allow for a greater number of occupied neutron states than proton states; hence $N > Z$. Thus the shell model provides a physical basis for the behavior previously illustrated in Figure 3.1 and for the inclusion of the symmetry term in the liquid drop model.

Problems

5.1. For a neutron in a three-dimensional infinite spherical well calculate the actual values of the energies of the 1s, 1p, 1d, and 2s states (in MeV). Construct a plot along the lines of Figure 5.12 illustrating the results.

5.2. (a) Find an expression for the energy levels of a three-dimensional harmonic oscillator. This should be described in most introductory quantum mechanics texts. Do not actually solve Schrödinger's equation for this problem. The three-dimensional harmonic oscillator potential is widely used as an approximation for the shell model potential, not because it is particularly accurate, but because it is simple and convenient.

(b) Draw an energy level diagram for these energy levels and determine the degeneracy of each level.

(c) Use this model to predict the magic numbers for the system.

5.3. Draw the ground state nucleon configuration for ^{19}F, ^{33}S, ^{55}Mn, and ^{91}Zr.

5.4. (a) From measured masses calculate the binding energy per nucleon for ^{38}K, ^{40}Ca, and ^{42}Sc.

(b) Discuss these results in terms of the semiempirical mass formula and the shell model.

5.5. Using an analogy with a particle in an infinite square well, explain the general trends that are observed in Figure 5.12.

6 CHAPTER | **Properties of the Nucleus**

6.1 GROUND STATE SPIN AND PARITY

In addition to properly predicting the nuclear magic numbers, the shell model is effective at describing other aspects of nuclear properties. Among these are nuclear spin and parity. These properties can be determined experimentally by looking at, for example, decay processes, transitions between nuclear states, and interactions of nuclei with magnetic fields. The spin of a nucleus is calculated from the shell model results given in the last chapter by considering the angular momenta of the individual nucleons. The total nuclear spin, denoted as $\vec{J}$, is the vector sum of the angular momenta of the individual nucleons, $\vec{j}$. Here uppercase characters are used to denote the overall property of the nucleus, while lowercase is used as in Chapter 5 to denote the properties of individual nucleons. Thus we write

$$\vec{J} = \sum \vec{j}_i. \qquad (6.1)$$

It is important to consider how the individual $\vec{j}_i$ align vectorially. This is analogous to the problem in solid state physics where the relative orientation of the electron spins in the various energy levels must be determined in order to calculate magnetic moments. In nuclear physics the situation is actually somewhat simpler because of the strong pairing tendency for nucleons. Of course we have to consider the properties of neutrons and protons independently and determine the total spin as

$$\vec{J} = \vec{J}_n + \vec{J}_p. \qquad (6.2)$$

To understand how to interpret this equation, consider the ground state properties of some specific types of nuclei. Following the assumptions of the extreme single particle shell model, we note that like nucleons will pair with j antiparallel yielding a net zero contribution to J.

Even–Even Nuclei In the case of an even number of neutrons and an even number of protons both J_n and J_p will be zero yielding a net zero J for the nucleus. This has been found to be true experimentally.

Even–Odd and Odd–Even Nuclei These are perhaps the most interesting in the context of the shell model and can be described by the so-called single nucleon model. The even species of nucleon will contribute zero to the overall J and the contribution of the odd species of nucleon will be determined by the properties of a single unpaired nucleon.

Odd–Odd Nuclei These are the most complex to deal with because both J_n and J_p are nonzero and the relationship between these two vector quantities is not, in general, known. This allows us to place certain limits on the magnitude of the resulting total J, since the minimum and maximum values will be determined for antiparallel and parallel alignment of J_n and J_p, respectively, as

$$\left| \vec{J}_n - \vec{J}_p \right| \leq \left| \vec{J} \right| \leq \left| \vec{J}_n + \vec{J}_p \right|. \tag{6.3}$$

The above rules can be used to predict (or at least understand) the behavior of particular nuclei. Before looking at some examples consider the question of parity. The parity of a nuclear state is given by the parity of the nuclear wave function. This can be either even (denoted +) or odd (denoted −) depending on the mathematical properties of the wave function. An even wave function is symmetric and has the property

$$\psi(\vec{r}) = \psi(-\vec{r}) \tag{6.4}$$

and an odd wave function is antisymmetric and has the property

$$\psi(\vec{r}) = -\psi(-\vec{r}). \tag{6.5}$$

For the single nucleon model the parity of the overall wave function is given by the parity of the wave function of the unpaired nucleon. The symmetry of the wave function for a given value of l defines the parity as

$$\pi = (-1)^l \tag{6.6}$$

where $l = 0, 1, 2, 3, \ldots$ correspond to s, p, d, f, ... states. Thus s-states have even parity, p-states have odd parity, d-states have even parity, and so on.

In standard spectroscopic notation the occupancy of each state is given by a superscript. As an example consider the odd A nuclei ^{15}O and ^{17}O. Both nuclides have a filled proton shell ($Z = 8$). ^{15}O has $N = 7$ while ^{17}O has $N = 9$. The state occupancy for ^{15}O and ^{17}O is given in Table 6.1 and is determined by

Table 6.1 | Nucleon Configurations for ^{15}O and ^{17}O

Nuclide	Proton State	Neutron State	J^π
^{15}O	$(1s_{1/2})^2(1p_{3/2})^4(1p_{1/2})^2$	$(1s_{1/2})^2(1p_{3/2})^4(1p_{1/2})^1$	$1/2^-$
^{17}O	$(1s_{1/2})^2(1p_{3/2})^4(1p_{1/2})^2$	$(1s_{1/2})^2(1p_{3/2})^4(1p_{1/2})^2(1d_{5/2})^1$	$5/2^+$

sequentially filling states in order of increasing energy. Figure 5.11 provides some guidance in this respect. In both cases the proton states contribute zero to J and the net nuclear angular momentum is the result of the properties of the one unpaired neutron. For ^{15}O the unpaired neutron is in a $1p_{1/2}$ state giving $J = J_n = 1/2$ and an odd parity as appropriate for a p-state wave function. This state is sometimes referred to as $J^\pi = 1/2^-$. For ^{17}O the unpaired neutron is in a $1d_{5/2}$ state giving $J = J_n = 5/2$ and even parity as appropriate for a d-state wave function. In many cases, particularly heavier nuclides, the spectroscopic notation is given by only the outermost level or levels (since inner filled levels do not contribute to the net nuclear spin or parity). Thus the neutron states for ^{15}O would be given as $(1p_{1/2})^1$ and for ^{17}O as $(1d_{5/2})^1$.

A slightly more complex example is provided by two isotopes of Si as shown in Table 6.2. Since both ^{27}Si and ^{29}Si have $Z = 14$ there is no contribution to the nuclear spin from the protons. The neutron configuration of ^{27}Si is $(1d_{5/2})^5$ indicating an unpaired d-state neutron giving a spin $5/2$ and even parity. For ^{29}Si a strict interpretation of Figure 5.11 would provide the incorrect spin and parity of $3/2^+$ (for the unpaired neutron). The correct interpretation of these states is given in Figure 5.12 where it is seen that the curves for $2s_{1/2}$ and $1d_{5/2}$ cross and for light nuclei the $2s_{1/2}$ state is filled before the $1d_{5/2}$. This gives the correct spin and parity of $1/2^+$. A careful examination of this figure shows that several such cases exist and provides a warning for over-interpreting a single energy level diagram such as Figure 5.11.

In general it is seen from a comparison of experimental results and calculations based on the single nucleon model that the ground state of even–even nuclei are properly described as 0^+. There is remarkable agreement between model and experiment for the ground state properties of odd A nuclei, provided a careful consideration of the ordering of the energy levels is taken. However, the model

Table 6.2 | Nucleon Configurations for ^{27}Si and ^{29}Si ·

Nuclide	Proton State	Neutron State	J^π
^{27}Si	$(1s_{1/2})^2(1p_{3/2})^4(1p_{1/2})^2(1d_{5/2})^6$	$(1s_{1/2})^2(1p_{3/2})^4(1p_{1/2})^2(1d_{5/2})^5$	$5/2^+$
^{29}Si	$(1s_{1/2})^2(1p_{3/2})^4(1p_{1/2})^2(1d_{5/2})^6$	$(1s_{1/2})^2(1p_{3/2})^4(1p_{1/2})^2(1d_{5/2})^6(2s_{1/2})^1$	$1/2^+$

fails to provide any simple systematic method of predicting the properties of odd–odd nuclei.

6.2 EXCITED NUCLEAR STATES

Nuclei can exist in excited states and these states are typically populated by decays from other unstable nuclides or by interactions with other particles. Since the typical spacing of excited states in a nucleus is of the order of an MeV, thermal excitations play a negligible role (unlike the case for electronic energy levels). Here we discuss excited states in the context of the shell model. Further information about excited states will be given in later chapters.

The properties of many excited nuclear states can be described by non-ground state distributions of nucleons in the energy levels. A reliable test for the accuracy of this kind of model is a direct comparison with experimentally determined nuclear spins and parities. The simplest description that is applicable to many excited nuclear states is the single particle model where an excited state is described by the excitation of a single nucleon. This approach is most applicable to nuclei that have one unpaired nucleon outside filled neutron and proton shells. This is easiest to see in an example. Here we consider ^{41}Ca, which has a filled proton shell and one unpaired $1f_{7/2}$ neutron as indicated in Figure 6.1. The properties of the ground state and the energy of the first few excited states of ^{41}Ca are illustrated in Figure 6.2. Most of these states can be described as single particle states as illustrated in Figure 6.3. For each single particle state a single

Figure 6.1 | Neutron and Proton Configuration for the Ground State of ^{41}Ca

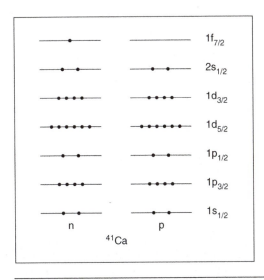

Figure 6.2 | Spins and Parities of the Energy Levels of ^{41}Ca

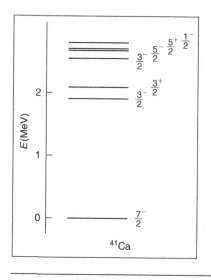

Figure 6.3 | Odd Neutron Configurations for the Single Nucleon Excited States of ^{41}Ca

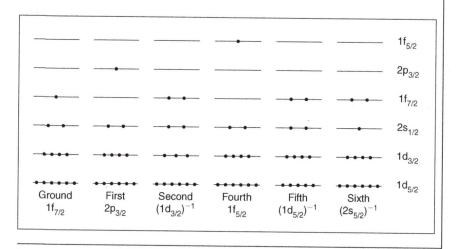

neutron is excited into a higher energy state resulting in an unpaired neutron in one of the states. This unpaired neutron readily gives the resulting overall nuclear spin and parity. The state designation as given in Figure 6.3 provides the relevant nuclear properties. In cases where a nucleon from a filled level is excited to a higher level resulting in an unpaired nucleon in a level that had previously been filled, this missing nucleon can be thought of as a "hole." The spectroscopic notation can then be used to denote the state in terms of this missing nucleon. It is therefore customary to refer to the second excited state of ^{41}Ca as $(1d_{3/2})^{-1}$ rather than $(1d_{3/2})^3$. Similar designations apply for the fifth and sixth excited states.

The shell model therefore provides a reliable means of determining the properties of excited states, in particular in the cases where excited state properties are accurately described by single particle states. In some cases an excited state cannot be described by the excitation of a single nucleon and two or more nucleons are involved giving rise to two or more unpaired nucleons. This may involve either neutrons or protons or both. The description of these states is analogous to the description of odd–odd nuclei and clearly defined rules cannot be described.

6.3 MIRROR NUCLEI

Mirror nuclei are pairs of nuclei that have the same values of A but the values of N and Z interchanged. Some examples are ^{3}H – ^{3}He, ^{7}Li – ^{7}Be, ^{9}Be – ^{9}B, ^{11}B – ^{11}C, ^{13}C – ^{13}N, ... The shell model would predict very similar properties for these pairs of nuclides and an experimental investigation of excited states in mirror nuclei gives some interesting insight. For one nuclide of each pair, for example, ^{13}C, the nuclear properties should be determined by a single unpaired neutron, for the other nuclide, for example, ^{13}N, the properties are determined by a single unpaired proton. If we ignore the presence of coulombic interactions then these cases are indistinguishable. Thus differences between the properties of mirror nuclei are an indication of the importance of coulombic interactions and similarities are suggestive of the importance of the strong interaction. The first few excited states of ^{13}C and ^{13}N are compared in Figure 6.4. The spins and parities of the energy levels shown in Figure 6.4 are the same, indicating that for the first few energy levels (at least) the single particle excitations are the same for the unpaired neutron in ^{13}C and the unpaired proton in ^{13}N.

6.4 ELECTROMAGNETIC MOMENTS OF THE NUCLEUS

The measurement of the electromagnetic moments of nuclei provides important information about nuclear structure, nuclear charge distributions, and nuclear shapes, as well as insight into the properties of the neutron and proton themselves. The electromagnetic moments of a nucleus result from the distribution of charges and currents within the nucleus. The electric monopole moment results from the total electric charge. In the case of a nucleus this is merely $+eZ$. For symmetry reasons the electric dipole moment of the nucleus (as well as all other static multipole moments with odd parity; for example, magnetic monopole, magnetic quadrupole, or electric octupole) must vanish. The electric quadrupole moment exists for charge distributions that are nonspherical. In previous chapters we assumed that nuclei are spherical. However, this is not always

Figure 6.4 | Excited States of the Mirror Nuclei ^{13}C and ^{13}N

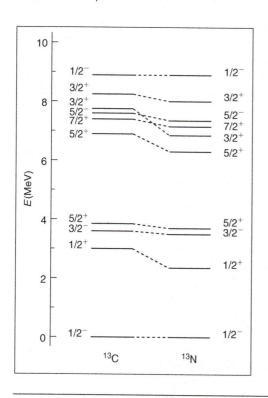

the case and a measure of the nuclear electric quadrupole moment is an indication of nonspherical nuclear symmetry. The magnetic multipole moment that is of interest is the magnetic dipole moment. This results from current distributions in the nucleus; recall the classical magnetic dipole consisting of a single current loop. Table 6.3 summarizes information about the lower order moments

Table 6.3 | Characteristics of the Lowest Order Electric and Magnetic Multipole Moments of the Nucleus

Moment	Electric	Magnetic
monopole	net nuclear charge	0
dipole	~0	due to currents
quadrupole	due to nonspherical $\rho_p(r)$	~0

of nuclei. The electric quadrupole moment and the magnetic dipole moment of the nucleus are of most relevance to the models described in previous chapters and these are discussed in detail below.

6.5 ELECTRIC QUADRUPOLE MOMENTS

The interaction energy of a charge distribution $\rho(r)$ and an external electric potential $\phi(r)$ can be written as

$$U = \int \rho(\vec{r})\phi(\vec{r})d^3\vec{r}. \tag{6.7}$$

If $\phi(r)$ is a slowly varying function of r then it can be expanded in a Taylor series about $r = 0$ as

$$\phi(\vec{r}) = \phi(0) + (\vec{r} \cdot \nabla \phi)_{r=0} + \frac{1}{2}\sum_{i,j} x_i x_j \left.\frac{\partial^2 \phi}{\partial x_i \partial x_j}\right|_{r=0} + \dots \tag{6.8}$$

where the x_i, and x_j are the spatial coordinates x, y, and z. Substituting this into equation (6.7) gives the interaction energy as

$$U = \phi(0)\int \rho(\vec{r})d^3\vec{r} + \left(\int \rho(\vec{r})\vec{r}\,d^3\vec{r}\right)\cdot\nabla\phi + \frac{1}{2}\sum_{i,j}\frac{\partial^2\phi}{\partial x_i \partial x_j}\int \rho(\vec{r})x_i x_j\,d^3\vec{r} + \dots \tag{6.9}$$

The first term on the right-hand side is the monopole term and defines the electric monopole moment (or total charge) as

$$eZ = \int \rho(\vec{r})d^3\vec{r}. \tag{6.10}$$

The second term on the right-hand side gives the electric dipole moment, P, as

$$P = \int \rho(\vec{r})\vec{r}\,d^3\vec{r}. \tag{6.11}$$

The third term on the right-hand side of equation (6.9) comes from the electric quadrupole moment of the nucleus and, in the case where there is axial symmetry along the z-axis, this gives the quadrupole moment, Q, as

$$Q = \int \rho(\vec{r})[3z^2 - r^2]d^3\vec{r}. \tag{6.12}$$

Some measured electric quadrupole moments are illustrated in Figure 6.5. An interesting feature of these data is the crossing from positive to negative values of Q for Z equal to a magic number. This feature is also observed for N equal to a magic number. In the context of the shell model it is encouraging to see that nuclei with filled shells have near zero Q and are, therefore, spherically symmetric.

Figure 6.5 | Electric Quadrupole Moments for Some Odd *A* Nuclei as a Function of *Z*

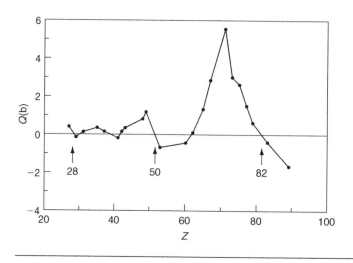

The meaning of nonzero Q can be readily understood from an inspection of equation (6.12). Prolate nuclei are those that are elongated along the z-axis and will have $Q > 0$. This is seen from the equation by writing $r^2 = x^2 + y^2 + z^2$ and observing that when one integrates over the volume of the nucleus, $3z^2$ will, on the average be greater than r^2. Oblate nuclei are compressed along the z-axis and, as equation (6.12) indicates, will have $Q < 0$. In a qualitative sense the shell model can explain positive Q values for nuclei with Z slightly less than a magic number in terms of the existence of proton hole states. Similarly a negative Q for Z greater than a magic number corresponds to unpaired proton states. Figure 6.6 shows that a nucleus with one less proton than a filled shell is approximated by a prolate ellipsoid and yields a positive quadrupole moment. The figure also shows how a nucleus with one proton more than a spherical closed shell can be approximated by an oblate ellipsoid and will give rise to a negative Q. Although the qualitative predictions of the shell model can explain the sign of the quadrupole moments, this model is rather ineffective at predicting the magnitude of Q. Figure 6.5 shows that nuclei without closed shells can have quite large nonspherical deformations. In the extreme cases this corresponds to $\Delta R/R$ of about 30% and this substantially exceeds expectations based on the single nucleon shell model. The question of neutron number is also difficult to deal with. Since the electric quadrupole moment as given by equation (6.12) involves the distribution of charges, the role of unpaired neutrons in determining Q is not obvious. These features of electric quadrupole moments of nuclei are discussed in Section 6.7 in terms of the collective model.

Figure 6.6 | Relationship of Proton and Hole States to Oblate and Prolate Elliptical Distortion

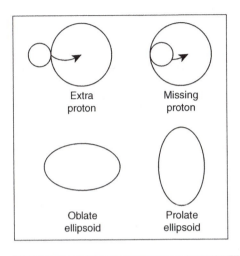

Extra
proton

Missing
proton

Oblate
ellipsoid

Prolate
ellipsoid

6.6 MAGNETIC DIPOLE MOMENTS

In the context of the shell model, the magnetic dipole moment of a nucleus is related to the properties of unpaired nucleons. Even–even nuclei are straightforward to deal with as the net nuclear spin is zero and the corresponding net dipole moment is also zero. This is observed experimentally. Odd–odd nuclei are difficult to understand because of the uncertainty in the relationship of the angular momentum of the odd neutron and that of the odd proton. Therefore, the most interesting cases for comparison with experimental observations are the odd A nuclei. Here it is necessary to consider the properties of the single unpaired nucleon. The nuclear dipole moment results from the total angular momentum of this nucleon. The components of the total angular momentum, that is, the spin and the orbital angular moments, can both contribute to the total dipole moment and the way in which these two components add must be considered in detail. Without worrying too much about the difference between even–odd and odd–even nuclei we begin with a fairly general discussion of a "generic" unpaired nucleon.

The nuclear magnetic dipole moment that arises from the orbital angular momentum, l, of an unpaired nucleon is given by

$$\vec{\mu}_l = \frac{g_l \mu_N \vec{l}}{\hbar} \qquad (6.13)$$

where μ_N is the nuclear magneton and is defined as $e\hbar/2m_p$ (m_p = proton mass). g_l is a quantity analogous to the Landé g-factor used in atomic physics and relates the magnitude of the dipole moment to the orbital angular momentum in units of $\hbar$. An analogous expression can be written for the spin component:

$$\vec{\mu}_s = \frac{g_s \mu_N \vec{s}}{\hbar}. \tag{6.14}$$

It is important to note that, although the vector sum of $\vec{s}$ and $\vec{l}$ is along the direction of $\vec{j}$ (equation (5.12)), the vector sum of the spin and orbital components of $\vec{\mu}$ as given by equations (6.13) and (6.14) is not along the direction of $\vec{j}$. The direction of $\vec{j}$ is well defined while the directions of $\vec{s}$ and $\vec{l}$ precess around $\vec{j}$. Similarly the direction of $\vec{\mu}$ precesses around $\vec{j}$ and it is the j component of $\vec{\mu}$, defined as μ, that is a measurable quantity. From equations (6.13) and (6.14) we define this component of μ as

$$\mu = \frac{g_l \mu_N}{\hbar} \frac{\vec{l} \cdot \vec{j}}{j} + \frac{g_s \mu_N}{\hbar} \frac{\vec{s} \cdot \vec{j}}{j}. \tag{6.15}$$

The first dot product on the right-hand side of this equation is found from the scalar product

$$s^2 = l^2 + j^2 - 2\vec{l} \cdot \vec{j} \tag{6.16}$$

to be

$$\frac{\langle \vec{l} \cdot \vec{j} \rangle}{\langle j \rangle} = \hbar \frac{l(l+1) - s(s+1) + j(j+1)}{2[j(j+1)]^{1/2}}. \tag{6.17}$$

Similarly the second dot product is obtained from

$$l^2 = s^2 + j^2 - 2\vec{s} \cdot \vec{j} \tag{6.18}$$

to be

$$\frac{\langle \vec{s} \cdot \vec{j} \rangle}{\langle j \rangle} = \hbar \frac{s(s+1) - l(l+1) + j(j+1)}{2[j(j+1)]^{1/2}}. \tag{6.19}$$

Combining equations (6.15), (6.17), and (6.19) gives

$$\mu = \frac{g_l \mu_N}{2} \left[\frac{l(l+1) - s(s+1) + j(j+1)}{[j(j+1)]^{1/2}} \right] + \frac{g_s \mu_N}{2} \left[\frac{s(s+1) - l(l+1) + j(j+1)}{[j(j+1)]^{1/2}} \right]. \tag{6.20}$$

Experimentally it is the component of μ along the direction of the magnetic field that is measured. This is obtained by multiplying equation (6.21) by a factor $j/[j(j+1)]^{1/2}$. After some simplification this yields

$$\mu = \frac{\mu_N}{2} \left[(g_l + g_s)j + (g_l - g_s) \frac{l(l+1) - s(s+1)}{j+1} \right]. \tag{6.21}$$

For protons and neutrons $s = \frac{1}{2}$ and, for a nuclear state defined by the properties of a single unpaired nucleon, the minimum and maximum allowed values of j are given by equation (5.13); that is $j = l \pm \frac{1}{2}$. Substituting these two values values for j into equation (6.21) yields

$$\mu = \frac{\mu_N}{2}[g_s + (2j - 1)g_l] \quad \text{for} \quad j = l + \frac{1}{2} \tag{6.22}$$

and

$$\mu = \frac{\mu_N}{2}\frac{j}{(j+1)}[-g_s + (2j + 3)g_l] \quad \text{for} \quad j = l - \frac{1}{2}. \tag{6.23}$$

It is now possible to apply the shell model to the prediction of nuclear magnetic dipole moments of specific nuclei with odd A. To do this we need to know the l and j values of the unpaired nucleon; this will tell us if we are dealing with the $j = l + \frac{1}{2}$ or the $j = l - \frac{1}{2}$ case. We also need to know the appropriate values for g_l and g_s. If we are considering even–odd nuclei then the values of g_l and g_s should be those appropriate for the unpaired proton. If we are considering odd–even nuclei then the g_l and g_s values should be those appropriate for the unpaired neutron. Since the neutron is uncharged its value of g_l is, by definition, zero. If the dipole moment in equation (6.13) is measured in units of the nuclear magnetons then the value of g_l for the proton is 1. Experimentally determined values of g_s for a free neutron and a free proton are given in Table 6.4. The nonzero g_s value for the uncharged neutron results from the fact that hadrons are not fundamental particles but have internal structure. This is discussed further in Chapter 15.

Equations (6.22) and (6.23) can readily be applied to the two cases for nuclei with odd A; those with odd N (an unpaired neutron) and those with odd Z (an unpaired proton). Values of μ in nuclear magnetons can be calculated as a function of j of the unpaired nucleon, for both neutron states and proton states and for both alignments of l and s. These values determine the so-called Schmidt lines as a function of j as illustrated in Figures 6.7 and 6.8. The data illustrated in the figures show that the model provides somewhat less than ideal agreement with experiment. In fact the Schmidt lines seem to be limiting values, although, in general, data points lie closer to the correct line. A number of possible explanations have been suggested for the lack of agreement observed here. Two possibilities as described below are of particular interest.

Table 6.4 | *g*-Factors for the Neutron and Proton

Nucleon	g_l	g_s
neutron	0	−3.8261
proton	1	5.5856

Figure 6.7 | Magnetic Dipole Moments and Schmidt Lines for Odd *N* Nuclei (Neutron States)

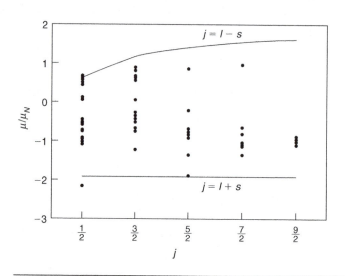

Figure 6.8 | Magnetic Dipole Moments and Schmidt Lines for Odd *Z* Nuclei (Proton States)

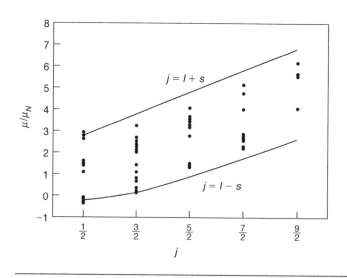

In many cases nuclear states may not be pure states but may be a mixture of states resulting from different configurations of neutrons and protons in the various energy levels, each of which yields the same spin and parity. It is convenient to think of states as defined by their wave functions and the resulting wave function being a combination of wave functions from states of equivalent spin and parity. This situation is referred to as configurational mixing and its inclusion in models of nuclear properties has been beneficial in describing electric quadrupole and magnetic dipole moments of nuclei.

The values of g_s that have been used for the neutron and proton are the values measured experimentally for free particles. It is interesting to note that these are both quite different than the value of 2.0, which is expected for point charges and that has been observed for the electron. In fact, the non–zero value of g_s for the uncharged neutron indicates that the internal structure of these particles plays an important role in determining g_s. It is not necessarily reasonable to assume that the values of g_s for nucleons in a nucleus should be the same as the free values. On the average, calculated moments show the best agreement with experimental results if g_s values for the neutron and proton are taken to be about 60% of their free values.

The points described above improve the agreement between calculated and measured nuclear properties but these (and some similar approaches) ultimately fall short of providing an ideal nuclear model. Another possible problem lies with the form of the nuclear potential that has been used in the shell model. This is taken to be spherically symmetric. It is known that nuclei with filled shells are reasonably spherical but from Figure 6.5 it is clear that nuclei with partially filled shells can show substantial nonspherical distortion.

6.7 AN OVERVIEW OF THE COLLECTIVE MODEL

Both the liquid drop model and the shell model have met with some success in describing the properties of nuclei. The liquid drop model ignores the quantum mechanical properties of the individual nucleons and is therefore unable to make predictions of nuclear properties such as spin and parity. This model has, however, been successful in describing properties such as total binding energy. The shell model has considered the properties of the individual nucleons and in particular the behavior of the unpaired nucleon in odd A nuclei. This approach has been surprisingly successful in determining nuclear spins and parities and in describing the properties of many excited states. Although the shell model correctly predicts the change in sign of the quadrupole moment near filled shells, it is not appropriate for explaining the very large quadrupole moments for some nuclei and even underestimates the small moments of nuclei with single particle states. It has also been seen that magnetic dipole moment calculations provide order of magnitude agreement with experiment but fail to give accurate numerical values. It would thus seem that the two models, which take quite different approaches, each have their own strengths and weaknesses. The collective model has been an attempt to

Figure 6.9 | Low Lying Excited States of ^{38}Ar

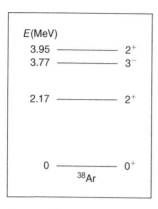

reconcile these two different approaches to better describe certain nuclear properties. The discussion here gives a brief overview of this model.

The basic idea of the collective model is to consider the filled shells of the nucleus as a central core that is described in terms of the liquid drop-like behavior of the nucleons. The quantum mechanical properties of the nucleus are described, as in the shell model, by the spins of the surface nucleons outside of the core. The motion of the surface nucleons introduces a nonsphericity to the central core and, in a practical sense, this type of behavior can be considered by a shell model with a nonspherically symmetric potential. This nonsphericity can account for the anomalous electric quadrupole moments that have been measured. It is important in such cases to realize that nonspherical nuclei can have energy associated with their rotational and/or vibrational degrees of freedom. In order to appreciate the importance of collective effects it is of interest to consider the excited states of some even–even nuclei. According to the shell model the ground states of even–even nuclei should all be 0$^+$. This is observed to be the case. We can consider the formation of excited states by the excitation of nucleons. The situation here is not so straightforward because, if a single nucleon is excited, then two levels will have unpaired nucleons that will contribute to the overall nuclear spin. From the shell model we have no guidelines for how to add these two contributions vectorially, although we can put limits on the resulting J value. An example helps to see some general features of even–even nuclei.

Some of the low-lying excited states of ^{38}Ar are illustrated in Figure 6.9. The neutron ($N = 20$) and proton ($Z = 18$) configurations for ^{38}Ar are shown in Figure 6.10. Since the neutron shell is filled, we expect that the low-lying excited states should be described by proton excitations. The simplest excitation would be an excitation of one of the $1d_{3/2}$ protons into the $1f_{7/2}$ state. This would give

Figure 6.10 | Ground State Nucleon Configuration for ^{38}Ar

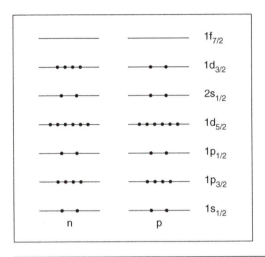

an integer spin between $7/2 - 3/2 = 2$ and $7/2 + 3/2 = 5$ depending on how the j vectors align. The parity of this state would be the product of the parities of $1d_{3/2}$ and $1f_{7/2}$ states; $(+1)(-1) = (-1)$. Figure 6.9 shows that this excitation probably corresponds to the 3^- state at 3.77 MeV. However, how can we explain the 2^+ state at 2.17 MeV? Another possible excitation would be a $2s_{1/2}$ proton into the $1d_{3/2}$ level resulting in $(2s_{1/2})^{-1}$ and $(1d_{3/2})^1$ states. This would give possible spin values of $3/2 - 1/2 = 1$ to $3/2 + 1/2 = 2$ and a parity $(+1)(+1) = (+1)$. This would explain the properties of the 2.17 MeV state. However, a careful inspection of (5.12) shows that this state should be at an energy that is very similar to the 3^- state. Thus this excitation nicely corresponds to the 2^+ state at 3.95 MeV. It is, in fact, difficult to explain the 2^+ state at 2.17 MeV. This is not an isolated case. In fact the first excited state of virtually all even–even nuclei are 2^+ states and these are difficult to explain on the basis of single nucleon excitations. Notable exceptions to this feature occur for doubly magic nuclei. The collective model is helpful in understanding the properties of the even–even nuclei and makes predictions on the basis of the collective motion of a large number of nucleons rather than a single unpaired nucleon.

In even–even nuclei that are not doubly magic the unfilled shells introduce a nonspherical distortion. Classically the energy associated with the rotation of an object with a moment of inertia I is given by

$$E = \frac{J^2}{2I} \tag{6.24}$$

Figure 6.11 | Rotational Levels in ^{174}W

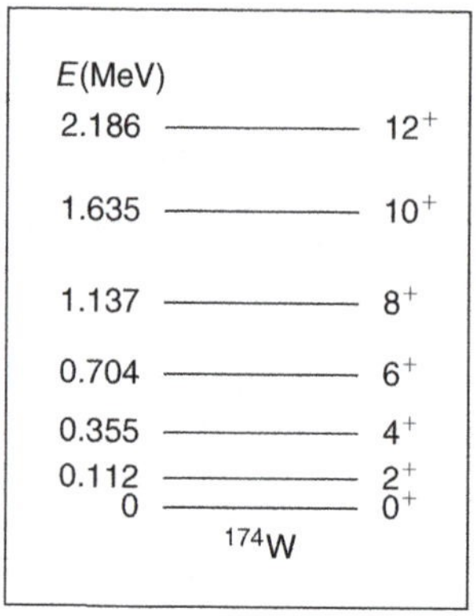

```
E(MeV)
2.186 ——————— 12⁺

1.635 ——————— 10⁺

1.137 ——————— 8⁺

0.704 ——————— 6⁺

0.355 ——————— 4⁺
0.112 ——————— 2⁺
   0  ——————— 0⁺
        ¹⁷⁴W
```

where J is the classical angular momentum. Quantum mechanically this is written in terms of the expectation values for J^2 as

$$E = \frac{J(J+1)\hbar^2}{2I}. \tag{6.25}$$

Because both N and Z are even, the nucleus will have a symmetric wave function and the values of J will be constrained to be even. Values of $J = +2, +4, +6, +8,\ldots$ will give rise to excited rotational states with spin and parity of 2^+, 4^+, 6^+, 8^+,.... The lowest lying 2^+ state accounts for the lowest energy excited state in many deformed even–even nuclei. (Another possibility for this 2^+ state is discussed below.) In light nuclei the higher order rotational states are usually intermixed with states resulting from nucleon excitations. In heavy nuclei, purely rotational energy levels are sometimes seen as illustrated in Figure 6.11 for ^{174}W. The relative spacing of the energy levels can be calculated from equation (6.25) and can be determined experimentally by normalizing measured energies to the energy of the 2^+ state. Table 6.5 shows that the predictions of the collective model are good for small J but tend to overestimate the energies for larger J. The energy of the first 2^+ excited state shows a generally decreasing trend with increasing nuclear mass. This can be explained in terms of equation (6.25). With increasing mass the corresponding increase in moment of inertia causes a reduction in the spacing of the rotational energy levels. The exceptions to this behavior occur for magic nuclei, especially those that are doubly magic where the high degree of spherical symmetry reduces the moment of inertia and the rotational modes are shifted to higher energy. The lowest lying states in these nuclei are the result of more complex behavior.

Other possible excitation modes for nuclei are the vibrational modes. Since nuclear matter is relatively incompressible, radial oscillations of the nucleus are

Table 6.5 | Measured and Calculated Excited State Energies (Relative to the 2$^+$ State) for the Rotational Modes of ^{174}W

Level	$E/E(2^+)$	
	Measured	Calculated
2$^+$	1.00	1.00
4$^+$	3.17	3.33
6$^+$	6.29	7.00
8$^+$	10.2	12.0
10$^+$	14.6	18.3
12$^+$	19.5	26.0

not typical. Rather the simplest kind of nuclear vibrations would be shape oscillations analogous to the shape oscillations experienced by a liquid drop. The lowest order vibrations of this kind are quadrupole vibrations where the nucleus would oscillate between a prolate ellipsoid and an oblate ellipsoid. These oscillations may be described using a harmonic oscillator potential. The Schrödinger equation gives the splitting of the quantum mechanical energy levels for this potential in terms of the oscillation frequency, ω, as

$$\Delta E = \hbar\omega. \qquad (6.26)$$

Such quantized nuclear vibrational states are referred to as phonons by analogy with quantized lattice vibrations in solid state physics. The $n = 1, 2, 3,...$ states have linearly increasing energy and are associated with 1, 2, 3,... phonons, respectively. The phonon has a spin and parity of $j^\pi = 2^+$ giving a spin and parity of the first excited vibrational state of 2$^+$. The spins of the two phonons associated with the second excited vibrational state can add vectorially to give a 0$^+$, 2$^+$, or 4$^+$ state. The three-phonon state corresponds to 0$^+$, 2$^+$, 3$^+$, 4$^+$, or 6$^+$ and so on for higher energies. In the ideal harmonic oscillator potential the energy states are degenerate in J but, in reality, perturbations that have not been discussed here lift the degeneracy and split each level into separate J states. An example of experimentally observed nuclear vibrational states is illustrated for ^{120}Te in Figure 6.12. This shows the singlet 2$^+$ first excited state and the triplet second excited state. For higher energy levels the situation becomes more complex and pure vibrational modes are not observed.

In most cases the description of nuclear energy levels is not as straightforward as in the simple examples above. Generally, nucleon excitations, rotational modes, and vibrational modes are not well decoupled. Resulting energy levels of excited states can be quite complex and cannot always be explained on the basis of model predictions. More sophisticated models have followed

Figure 6.12 | Vibrational Levels in ^{120}Te

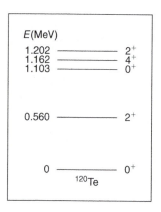

from the shell model and the collective model and have provided improved predictions of nuclear state properties. Some of these are discussed in the more advanced texts given in the bibliography.

Problems

6.1. Consider the first eight excited states of a ^{13}C nucleus. Describe possible nucleon states for as many of these states as possible.

6.2. (a) For the data in Figure 6.11 calculate the moment of inertia for a ^{174}W nucleus in each of the energy levels shown.
(b) Assuming that a ^{174}W nucleus is approximately spherical, calculate the classical moment of inertia of a sphere of appropriate diameter and mass. Compare with part (a).

6.3. Using the data in Figure 6.12, calculate the phonon frequency for the vibrational modes of a ^{120}Te nucleus.

6.4. (a) For the following odd–odd nuclei use the shell model to determine the expected nuclear parity and the range of possible values of the nuclear spin: ^{14}N, ^{20}F, ^{24}Na, and ^{26}Al.
(b) Compare the results of part (a) with the actual measured spins and parities.

6.5. Use the shell model to find the ground state spins and parities of ^{91}Y, ^{91}Zr, and ^{91}Nb.

6.6. Use the shell model to describe the spin and parity of the ground state ($3/2^-$) and first three excited states ($5/2^-$, $1/2^-$, and $3/2^-$, respectively, with increasing energy) of ^{59}Ni.

6.7. It is observed that odd A Sb nuclei with odd A have $J^\pi = 5/2^+$ for $A \le 121$ and $J^\pi = 7/2^+$ for $A \le 121$. Explain.

Nuclear Decays and Reactions

General Properties of Decay Processes

CHAPTER **7**

7.1 DECAY RATES AND LIFETIMES

In the past few chapters we have seen that certain nuclei are unstable and decay to more stable configurations of neutrons and protons. In Chapters 8 to 10 detailed accounts of the three most common nuclear decay processes will be given. Decay processes as related to fundamental particles will also be encountered in Section IV of this book. In this chapter some general properties of decays will be considered.

In a collection of identical unstable nuclei the number of decays per unit time will be proportional to the number of nuclei of that species that are present as a function of time, $N(t)$; that is,

$$-dN(t) = \lambda N(t)\,dt \qquad (7.1)$$

where the proportionality constant λ is the decay constant or decay rate. In general the nuclear species that decays is called the parent and the nuclear species that is produced from the decay is referred to as the daughter. Equation (7.1) is easily integrated to yield

$$N(t) = N(0)e^{-\lambda t} \qquad (7.2)$$

where $N(0)$ is the number of parent nuclei present at $t = 0$. The halflife of the decay process, $\tau_{1/2}$, is the time required for the initial number of nuclei to decay to one half. That is, substituting $N(t) = N(0)/2$ gives

$$\tau_{1/2} = \frac{\ln 2}{\lambda}. \qquad (7.3)$$

We can also define the mean lifetime, τ, which gives the mean time a nucleus survives in its initial state after creation. This is the integral of the decay time weighted by the decay rate:

$$\tau = \frac{\int_0^\infty \left(-\frac{dN(t)}{dt} \right) t\, dt}{\int_0^\infty \left(-\frac{dN(t)}{dt} \right) dt}. \tag{7.4}$$

Substituting equations (7.1) and (7.2) into the above and integrating gives

$$\tau = \frac{1}{\lambda} = \frac{\tau_{1/2}}{\ln 2}. \tag{7.5}$$

This expression shows the relationship between lifetime and halflife as these terms are used in this book.

Several different experimental techniques have been used for the measurement of nuclear lifetimes, depending on the time scales involved. For moderate to long lifetimes, most techniques involve the measurement of the decay rate. This is defined as the number of decays per unit time and is defined from equations (7.1) and (7.2) as

$$\left| \frac{dN(t)}{dt} \right| = \lambda N(0)e^{-\lambda t}. \tag{7.6}$$

For nuclei with very long lifetimes, λ is small (see equation (7.5)) and the exponential in equation (7.6) can be approximated as unity so that

$$\tau = \frac{1}{\lambda} = \frac{N(0)}{\left| dN(t)/dt \right|}. \tag{7.7}$$

Thus a knowledge of the number of nuclei present, $N(0)$ and a measure of the decay rate provides the lifetime. This method is suitable for very long lived nuclides, i.e. millions of years, and can be used for any lifetimes that are long compared to the time scale on which a measurement of $N(0)$ and $dN(t)/dt$ can be made, perhaps a few hours. A practical example is discussed in Section 7.3.

Moderate lifetimes can usually be obtained directly from equation (7.6) by successive measurements of the decay rate over a time period comparable to the lifetime. Taking the natural logarithm of both sides of equation (7.6) gives

$$\ln \left| \frac{dN(t)}{dt} \right| = \ln \left| \frac{dN(t)}{dt} \right|_{t=0} - \lambda t \tag{7.8}$$

where we have written

$$\lambda N(0) = \left| \frac{dN(t)}{dt} \right|_{t=0}. \tag{7.9}$$

Figure 7.1 | Results of a Counting Experiment to Determine the Lifetime of ^{64}Cu

The slope of the line gives the lifetime from equation (7.8) as 18.6 hours.

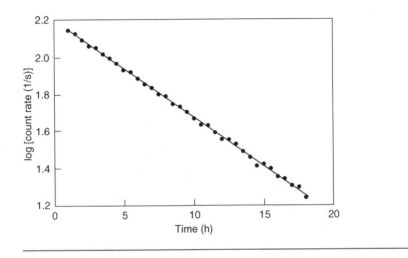

Figure 7.1 shows some experimental data for the decay of ^{64}Cu. The slope of the semilog plot yields the lifetime from equation (7.8). The lifetime of this nuclide is ideal for the application of this technique. Shorter lifetimes can certainly be measured, particularly with the use of appropriate computer-controlled instrumentation to make rapid successive measurements of $dN(t)/dt$. Lifetimes shorter than about 10^{-3} seconds require the use of techniques that, in many cases, depend on the nature of the decay processes and the method by which the parent nuclide is produced. These techniques can be roughly divided into two categories, direct and indirect measurements.

Direct measurements measure the time interval between the formation of the parent state and its decay to the daughter state. One example is the so-called coincidence technique. In many cases the parent state may be populated by the decay from another state and a particle (often a γ-ray) may be emitted during this process. The subsequent decay of the parent to the daughter may emit another particle and the time interval between these two events can be measured electronically.

Various indirect methods have been utilized for measuring lifetimes as short as 10^{-20} seconds and many are described in the nuclear instrumentation texts given in the bibliography. In general, these methods measure a quantity that can by appropriate theoretical considerations be related to the lifetime. These methods are not discussed in detail in this book, but much of the relevant background is presented in Sections 7.2 and 11.5.

Figure 7.2 | Multimodal Decay of ^{164}Ho

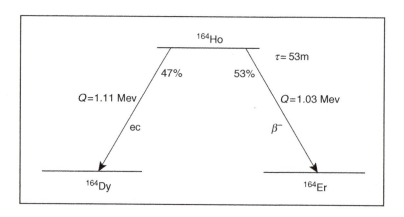

In many cases a single decay process between a parent state and a daughter state does not occur. Two more complex situations, multimodal decays and sequential decays, can also be considered. An example of a multimodal decay is illustrated in Figure 7.2. Here ^{164}Ho is an odd–odd nucleus similar to that shown in Figure 4.5 and may decay by either of two modes; β^- decay to ^{164}Er or electron capture to ^{164}Dy (see Chapter 9 for further information about these processes). The former occurs for 53% of the ^{164}Ho nuclei and the latter for the remaining 47%. These fractions are referred to as the branching ratios (or when written as 0.47 and 0.53, as the branching fractions) of the decay and each of these processes is referred to as a decay mode. This decay process can be characterized by two decay constants, λ_1 and λ_2, referred to as partial decay constants, which correspond to the probability of decay by β^- and electron capture, respectively. Analogous to equation (7.1) we can write

$$\frac{dN(t)}{dt} = -[\lambda_1 N(t) + \lambda_2 N(t)]$$ (7.10)

and

$$N(t) = N(0)e^{-(\lambda_1 + \lambda_2)t}$$ (7.11)

where a total decay constant, λ, is defined as the sum

$$\lambda = \lambda_1 + \lambda_2.$$ (7.12)

From equation (7.5) this gives the total lifetime as

$$\tau = \frac{1}{\lambda_1 + \lambda_2}.$$ (7.13)

Figure 7.3 | The Sequential Decay ^{218}Rn $\rightarrow$ ^{214}Po $\rightarrow$ ^{210}Pb

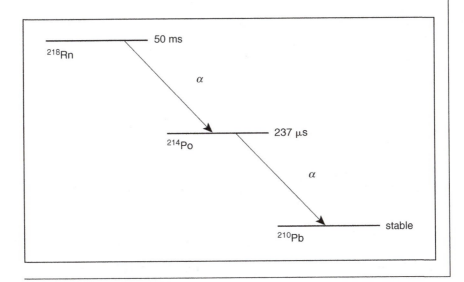

The branching fractions are given by the ratios of decay constants as

$$f_1 = \frac{\lambda_1}{\lambda} \quad \text{and} \quad f_2 = \frac{\lambda_2}{\lambda}. \tag{7.14}$$

These derivations can be generalized for multimodal decays with more than two modes by replacing equation (7.12) with a sum of the partial decay constants for all modes.

Sequential decays occur when the daughter of one decay process is unstable and subsequently decays. We have seen such a situation in Figure 4.3. A typical example of a sequential decay is illustrated in Figure 7.3. This shows the α-decay of ^{218}Ru (the parent) to ^{214}Po (the daughter), which then α-decays to stable ^{210}Pb (the granddaughter). Further details of the α-decay process will be discussed in Chapter 8. The lifetimes for these two decays as given in the figure are related to corresponding decay constants, λ_1 and λ_2. The relevant differential equations for this process can be written as follows: The number of parent nuclei, N_1, decreases as the parent decays:

$$dN_1(t) = -\lambda_1 N_1(t)\,dt. \tag{7.15}$$

The change in the number of daughter nuclei can be written as the sum of the new nuclei formed by the decay of the parent and those that are lost due to decay to the granddaughter:

$$dN_2(t) = \lambda_1 N_1(t)\,dt - \lambda_2 N_2(t)\,dt. \tag{7.16}$$

The solution to equation (7.15) can be found by direct integration:

$$N_1(t) = N_1(0)e^{-\lambda_1 t} \tag{7.17}$$

Using this result, the solution to the first order differential equation (7.16) with the boundary condition that $N_2(0) = 0$ is found to be

$$N_2(t) = N_1(0)\frac{\lambda_1}{\lambda_2 - \lambda_1}(e^{-\lambda_1 t} - e^{-\lambda_2 t}). \tag{7.18}$$

The decay rates can be obtained from equations (7.17) and (7.18) by differentiation.

The development given above can be extended (with considerable increase in mathematical complexity) to systems involving a greater number of sequential decays.

7.2 QUANTUM MECHANICAL CONSIDERATIONS

In Chapter 5 we discussed the application of the Schrödinger equation to the description of nuclear states in terms of the spatial component of the nuclear wave function, $\Psi(r, t)$. If we consider the application of these ideas to the description of nuclear decay processes, it is obvious that the time development of the wave function is important. In general, the time dependent wave function can be expressed in terms of the stationary states, $\psi(r)$ as:

$$\Psi(r, t) = \psi(r)e^{-iE_0 t/\hbar} \tag{7.19}$$

where E_0 is the energy of the state. If the state is unstable then we must also consider the time dependence predicted by the radioactive decay law. Since the probability of finding the nucleus is given by the modulus square of the wave function, equation (7.2) requires that this probability decay with time as

$$\left|\Psi(r, t)\right|^2 = \left|\Psi(r, 0)\right|^2 e^{-\lambda t}. \tag{7.20}$$

This expression suggests that equation (7.19) can be modified to give

$$\Psi(r, t) = \psi(r)e^{-iE_0 t/\hbar - \lambda t/2}. \tag{7.21}$$

This expression implies that a time dependent state (one that is not absolutely stable) is not represented by a single value of the energy, E_0, but may be described by a distribution of energies $P(E)$, with a mean value E_0. To examine this energy dependent distribution, the time dependent form of the wave function in equation (7.21) is Fourier transformed to give

$$P(E) = (\text{constant}) \cdot \int_{-\infty}^{+\infty} e^{-iE_0 t/\hbar - \lambda t/2} e^{iEt/\hbar} \, dt \qquad (7.22)$$

where the constant is required to properly normalize $P(E)$. If the decay of the initial state begins at $t = 0$, then the lower limit of the integration equation (7.22) can be set to zero. This integration yields

$$P(E) = \frac{(\text{constant})}{i(E_0 - E) + \lambda\hbar/2}. \qquad (7.23)$$

Normalization of $P(E)$ in the above expressions gives

$$P(E) = \left(\frac{i}{2\pi}\right)\left(\frac{1}{(E - E_0) + i\hbar\lambda/2}\right). \qquad (7.24)$$

The probability of finding the nucleus with a particular value of the energy, E, is given by the modulus of $P(E)$ squared:

$$|P(E)|^2 = \frac{1}{4\pi^2} \frac{1}{(E - E_0)^2 + \dfrac{\hbar^2\lambda^2}{4}}. \qquad (7.25)$$

This function is a Lorentzian, as illustrated in Figure 7.4 and a simple inspection of equation (7.25) shows that this curve has a full width at half maximum, Γ, of

$$\Gamma = \hbar\lambda. \qquad (7.26)$$

This equation gives the width of the energy distribution of an unstable state, $\Delta E \sim \Gamma$, and can be written in more common terminology as

$$\Delta E \Delta t = \hbar \qquad (7.27)$$

where the uncertainty in time, Δt, is the lifetime, τ, of the unstable state.

These concepts will become important in later chapters. However, for the present we should emphasize that the widths of nuclear state energy distributions are really very small. If we consider lifetimes greater than 10^{-15}s then the corresponding widths given by equation (7.26) are less than 10^{-6} MeV. As decay energies are commonly in the range of 10^{-2} to 10^{-1} MeV or greater, we can see that for such lifetimes the states are still very well defined and it is suitable for most purposes to refer to the transitions as monoenergetic.

Figure 7.4 | Heisenberg Width of the Energy Distribution from a Decay Process

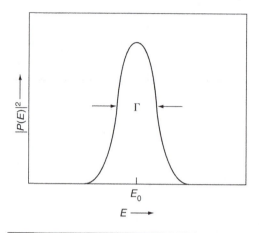

7.3 RADIOACTIVE DATING

An important application that makes use of the concepts described above is radioactive dating. This is a very useful technique for determining the age of organic systems or mineralogical samples. In the former case, ages of $10^4 - 10^5$ years can be determined using ^{14}C dating methods (lifetime = 8,270 years). In the latter case, various nuclides can be used for measuring ages up to a few billion years. Consider a simple example. A sample is produced at time $t = 0$, which contains a number of parent nuclei that decay to daughter nuclei with a known mean lifetime $\tau = 1/\lambda$. We make the following assumptions:

1. The sample contains no daughter nuclei at $t = 0$,
2. The daughter nuclei are not produced by any processes other than the decay of the parent.
3. The daughter nuclei are stable.

Thus we can write the number of parent nuclei, $N_1(t)$ and daughter nuclei $N_2(t)$ at time t (the present) as

$$N_1(t) + N_2(t) = N_1(0). \tag{7.28}$$

The radioactive decay law gives the time dependence of the number of parent nuclei as

$$N_1(t) = N_1(0)e^{-\lambda t}. \tag{7.29}$$

These equations can easily be combined to give t as

$$t = \frac{1}{\lambda} \ln\left[1 + \frac{N_2(t)}{N_1(t)}\right]. \tag{7.30}$$

Since the ratio $N_2(t)/N_1(t)$ and the decay constant, λ, can be measured in the laboratory the time between the formation of the sample and the present, t, can be calculated.

Unfortunately in most cases some quantity of the daughter nuclei may also be present in the sample at the time of formation. Thus equation (7.28) must be written as

$$N_1(t) + N_2(t) = N_1(0) + N_2(0). \tag{7.31}$$

This introduces an additional unknown into the equations above and does not allow for a direct determination of t from laboratory measurements. In many cases however, an additional, stable isotope of the daughter nuclide is also present in the sample at the time of formation. If additional quantities of this nuclide are not produced by any decay processes during the life of the sample, then this provides a means for determining the initial quantity of daughter nuclei in the sample. We refer to the number of nuclei of this stable isotope as N_s, and since this remains constant in time we can write

$$N_s(t) = N_s(0) \tag{7.32}$$

and equation (7.31) can be written as

$$\frac{N_1(t) + N_2(t)}{N_s(t)} = \frac{N_1(0) + N_2(0)}{N_s(0)}. \tag{7.33}$$

This can be combined with equation (7.29) for $N_1(0)$ to give

$$\frac{N_2(t)}{N_s(t)} = \frac{N_1(t)}{N_s(t)}[e^{\lambda t} - 1] + \frac{N_2(0)}{N_s(0)}. \tag{7.34}$$

$N_2(t)/N_s(t)$ and $N_1(t)/N_s(t)$ can be measured in the laboratory. Assuming that samples with a common origin should have the same value of $N_2(0)/N_s(0)$ (which is generally believed to be a good assumption), then equation (7.34) allows for the calculation of the age, t.

As an example of the above, consider the β^--decay $^{87}\text{Rb} \rightarrow {}^{87}\text{Sr}$, which has a lifetime of 6.9×10^{10} years. Another stable isotope of Sr, ^{86}Sr, also exists. So in the above equations $N_1(t)$ is the number of ^{87}Rb nuclei, $N_2(t)$ is the number of ^{87}Sr nuclei, and $N_s(t)$ is the number of ^{86}Sr nuclei as measured at the present time. Thus plotting $^{87}\text{Sr}/^{86}\text{Sr}$ as a function of $^{87}\text{Rb}/^{86}\text{Sr}$ for various samples with a common origin will yield a straight line with a slope of $(e^{\lambda t} - 1)$ from which the age, t, can be determined.

Figure 7.5 shows typical $^{87}\text{Sr}/^{86}\text{Sr}$ and $^{87}\text{Rb}/^{86}\text{Sr}$ ratios for some meteorite samples that are believed to be of common origin. These values show a straight line for $^{87}\text{Sr}/^{86}\text{Sr}$ as a function of $^{87}\text{Rb}/^{86}\text{Sr}$ where the slope yields an age for the five meteorites of 4.4×10^9 years. This value is consistent with ideas concerning the origin of meteorites and other estimates of the age of the solar system.

Figure 7.5 | The Ratio $^{87}Sr/^{86}Sr$ Plotted as a Function of $^{87}Rb/^{86}Sr$ for Some Meteorite Samples

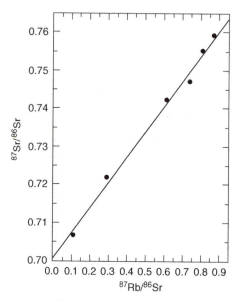

Problems

7.1. Find a suitable reference and learn about ^{14}C dating methods. Write a two- to three-page description of this method.

7.2. The human body contains about 20% carbon. Calculate the activity (in Curies) from ^{14}C for an average person.

7.3. Calculate the number of electrons emitted per second from a 1 gram sample of ^{137}Cs.

7.4. Natural Uranium, as found on earth, consists of two isotopes in the ratio of $^{235}U/^{238}U = 7.3 \times 10^{-3}$. Assuming that these two isotopes existed in equal amounts at the time the earth was formed, calculate the age of the earth. Note that the lifetimes of ^{235}U and ^{238}U are 1.03×10^9 years and 6.49×10^9 years, respectively.

7.5. ^{40}K decays by β^--decay with a branching ratio of 89%. One gram of natural potassium emits 27 electrons per second from this decay. Calculate the lifetime of ^{40}K.

7.6. Three radioactive sources each have an activity of 1 mCi at $t = 0$ and lifetimes of 1.0 m, 1.0 h, and 1.0 d. Calculate the decay rate for each source at $t = 1s$, 1m, 1h, and 1d.

7.7. Naturally occurring Vanadium contains 0.25% ^{50}V. This nuclide decays by β^--decay with an estimated lifetime of 3×10^{16} y. Calculate the number of electrons emitted per second from a 1 g sample of natural Vanadium as a result of the presence of this isotope.

7.8. Two radioactive nuclides, A and B, are present at $t = 0$ with the number of nuclei of A equal to twice the number of nuclei of B. The lifetimes of the two nuclides are τ for A and 3τ for B. Calculate the time (in units of τ) at which the number of nuclei of A and B are equal.

7.9. Consider the sequential decay where nuclide A decays to nuclide B with a lifetime τ_1, and nuclide B decays to nuclide C with a lifetime τ_2. If a sample contains only nuclide A at $t = 0$, calculate the time at which the decay rate of nuclide B is a maximum.

8 CHAPTER | Alpha Decay

8.1 ENERGETICS OF ALPHA DECAY

Alpha decay (α-decay) is the spontaneous emission of an α-particle, that is a ^{4}He nucleus or a bound system of two neutrons and two protons. The process can be described as

$$_Z^A X^N \rightarrow {}_{Z-2}^{A-4} Y^{N-2} + \alpha \tag{8.1}$$

and is known to occur in a few very light nuclides and many heavy nuclides. The energy release, Q, during α-emission can be calculated for the process in equation (8.1) as

$$Q = \left[m_N\left(_Z^A X^N\right) - m_N\left(_{Z-2}^{A-4} Y^{N-2}\right) - m_\alpha \right] c^2 \tag{8.2}$$

where the subscript N on the masses refers to nuclear masses. Since, in most cases it is convenient to deal with measured atomic masses, equation (8.2) can be written as

$$Q = \left[m\left(_Z^A X^N\right) - m\left(_{Z-2}^{A-4} Y^{N-2}\right) - m(^4\text{He}) \right] c^2 \tag{8.3}$$

where the electron masses are properly accounted for and the electronic binding energy has been ignored. It is assumed that the transitions are between nuclear ground states. Here it is important to realize that $m(^4\text{He})$ is the mass of a neutral ^{4}He atom and not the α-particle mass. If Q is negative then the process

Table 8.1	Energy, Q, Associated with the Emission of Various Particles form a ^{235}U Nucleus

Emitted Particle	Q (MeV)
n	−5.30
p	−6.70
^{2}H	−9.71
^{3}H	−9.97
^{3}He	−9.46
^{4}He	+4.68
^{6}Li	−3.85
^{7}Li	−2.88
^{7}Be	−3.79

is endothermic and cannot occur spontaneously. If Q is greater than zero then the process is exothermic and can occur spontaneously (at least from an energetic standpoint). In this case the excess energy is given up to the α-particle and the daughter nucleus in the form of kinetic energy. Because the number and identity of the nucleons do not change during α-decay, the nuclear binding energies can be directly substituted for the atomic masses in equation (8.3) (with an appropriate change of sign) as

$$Q = -\left[B\left(^{A}_{Z}X^{N}\right) - B\left(^{A-4}_{Z-2}Y^{N-2}\right) - B(^{4}\text{He})\right] \qquad (8.4)$$

where the binding energy of ^{4}He is known to be 28.3 MeV. Alpha emission is energetically favorable in many instances while emission of other light nuclei from a heavy nucleus is rather unlikely. The reason for this is that the binding energy of the ^{4}He is anomalously large (since it is a doubly magic nucleus). Table 8.1 shows the Q for emission of various light nuclei from a ^{235}U nucleus.

The value of Q for an α-decay process as given in equation (8.3) can be calculated on the basis of the measured atomic masses. These results are shown as a function of A in Figure 8.1. The following general features are seen in these data:

1. There is a crossing from negative Q to positive Q at around $A = 150$.
2. Local minima in the value of Q occur for some values of A.

These observations can be interpreted in the following terms. For A less than about 150 the process is endothermic and α-decay does not occur; for A greater than about 150 the process becomes exothermic and can, in principle, occur with a Q value that generally increases with increasing A. The minima near particular values of A are the result of shell effects and correspond directly to the features in the binding energy as seen in Figure 5.3.

Figure 8.1 | α-Decay Energies as a Function of A as Determined from Measured Atomic Masses

Data are shown for β-stable nuclides.

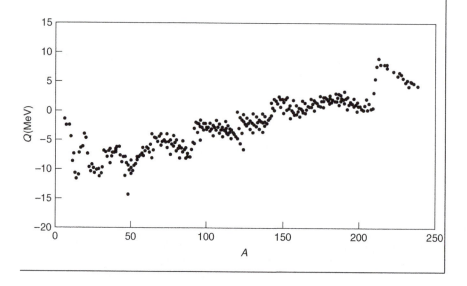

In general, experimental information about α-decay includes a measurement of the kinetic energy, T_α, of the α-particles that are released during the decay process. It is important to understand the relationship between this quantity and the value of Q discussed above. The energy that is liberated during the decay goes into kinetic energy that is distributed between the α-particle and the daughter nucleus. A simple consideration of conservation of energy and momentum shows that

$$T_\alpha = \frac{Q}{1 + \dfrac{m_\alpha}{m_D}} \tag{8.5}$$

where m_D is the mass of the daughter nucleus. For a typical α-decay process involving a heavy nucleus the daughter recoil energy accounts for about 2% of the total energy.

The Geiger-Nuttall rule states that there is a dramatic decrease in the α-decay lifetime with increasing decay energy. An example is illustrated in Figure 8.2. For even–even nuclei with a constant value of Z (for example, the various isotopes of Th) there is a smooth relationship between Q and τ as shown in the figure. The figure also shows that a similar relationship exists for a different value of Z, although data for different Z do not fall on the same line. Even–odd, odd–even, and odd–odd nuclei show the same general features, although for

Figure 8.2 | Geiger-Nuttall Relationship Between the α-Decay Halflife and the Decay Energy for Some Even Z Nuclei

Each line represents data for a different value of Z as indicated by the element name.

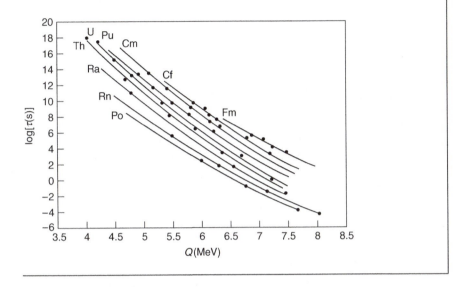

reasons described below there is some scatter introduced in the data if they are plotted on the same graph as the even–even nuclei. An interesting aspect of this figure is the range of values that are shown. For ^{232}Th with $Q = 4.08$ MeV the lifetime is 6×10^{17} seconds and for ^{218}Th with $Q = 9.85$ MeV the lifetime is 1.4×10^{-7} s. That is, a factor of about 2 in energy corresponds to a factor of 10^{24} in lifetime. This behavior explains the existence of many stable nuclei with A greater than 150 that have positive values of Q. For a value of Q less than about 4 MeV the lifetime becomes sufficiently long that the decay is, in practice, not observed. A description of the behavior illustrated by the Geiger-Nuttall relationship is one of the principal goals of the theory of α-decay presented below.

8.2 THEORY OF ALPHA DECAY

The basic theory of α-decay considers the probability that two neutrons and two protons will become bound together within a nucleus, thereby creating an α-particle and that this particle will then escape from the nucleus. The lifetime for α-decay will then be simply given in terms of the time scale for α-particle formation within the nucleus, τ_0, and the probability that the α-particle having been formed will escape from the nucleus, P, as

$$\tau = \frac{\tau_0}{P}. \qquad (8.6)$$

Figure 8.3 | Potential Well for the α-Decay Model

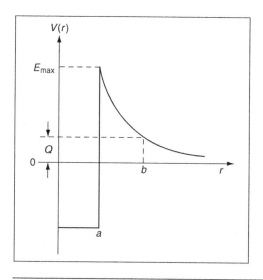

We will begin with a consideration of the escape probability, while the time τ_0 will be discussed briefly at the end of this section.

The behavior of the α-particle within the nucleus can be understood from a careful inspection of the nuclear potential well. Assuming that an α-particle has formed within the nucleus we must consider the behavior of this particle in the potential well of the daughter nucleus. This potential as illustrated in Figure 8.3 consists of the more-or-less spherical square well dominated by the strong interaction when the α-particle is inside the nucleus; that is, for $r < a$ where $a = R_D + R_\alpha$ (R_D is the radius of the daughter nucleus and R_α is the radius of the α-particle). Outside of the nucleus the potential is described by the coulombic interaction between the α-particle and the daughter nucleus:

$$V(r) = \frac{2Z_D e^2}{4\pi\varepsilon_0 r} \tag{8.7}$$

where the numerator gives the product of the charge of the α-particle $+2e$ and the daughter nucleus $+Z_D e$. This represents the coulombic barrier. If the Q for α-emission is positive then the α-particle will form inside the nucleus with an energy Q above zero potential as shown in the figure. Classically speaking the α-particle can escape only if Q is greater than the maximum height of the coulombic barrier, E_{max}. For a typical heavy nucleus, E_{max} is about 35 MeV. A comparison with the α-decay energies shown in Figure 8.1 demonstrates that this is never the case. The problem of α-decay then becomes a quantum mechanical problem of coulombic barrier penetration by the α-particle.

Inside the nucleus the potential can be approximated as a constant and the solutions for R, the radial part of the Schrödinger equation, follow along the lines described in Chapter 5. Outside the nucleus the radial part of the Schrödinger equation can be written as

$$-\frac{\hbar^2}{2m}\left[\frac{d^2R}{dr^2}+\frac{2}{r}\frac{dR}{dr}\right]+\left[\frac{2Z_D e^2}{4\pi\varepsilon_0 r}+\frac{l(l+1)\hbar^2}{2mr^2}\right]R=QR \qquad (8.8)$$

where the reduced mass has been defined in terms of the α-particle mass and the daughter nucleus mass, as

$$m=\frac{m_\alpha m_D}{m_\alpha+m_D}. \qquad (8.9)$$

We can define two regions outside the nucleus: $a<r<b$ and $r>b$ where b is defined as the radius at which $V(r)=Q$ or

$$b=\frac{2Z_D e^2}{4\pi\varepsilon_0 Q}. \qquad (8.10)$$

The radius a can be determined from equation (3.23). Solving equation (8.8) in these two regions is not simple. However, the inspection of a very simplified potential as shown in Figure 8.4 is helpful. This shows the penetration of a square barrier of height V_0 by the α-particle wave function. We also begin with

Figure 8.4 | Tunneling of an α-Particle Wave Function Through a Square Barrier

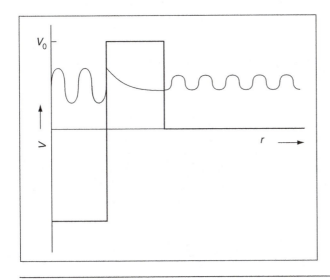

a consideration of the simple $l = 0$ case. Under these conditions the solutions to equation (8.8) are

$$R(r) = \frac{1}{r}\exp\left[\pm\sqrt{\frac{2mr^2}{\hbar^2}(V_0 - Q)}\right] \tag{8.11}$$

giving an exponentially damped solution within the barrier and an oscillatory solution outside the barrier. The barrier can be approximated by a series of square barriers of diminishing height as described by equation (8.7). This is the so-called WKB (Wentzel-Kramers-Brillouin) approximation and is discussed in detail in most introductory quantum mechanics texts. In the limiting case this series is written as an integral and (8.11) becomes

$$R(r) = \frac{1}{r}\exp\left(\pm\int F(r')dr'\right) \tag{8.12}$$

where we have defined the function

$$F(r') = \sqrt{\frac{2m}{\hbar^2}\left(\frac{2Z_De^2}{4\pi\varepsilon_0 r'} - Q\right)}. \tag{8.13}$$

Equation (8.12) can now be used to obtain the solution for $R(r)$ in the regions outside the nucleus. In principle we should obtain solutions for $r < a$, $a < r < b$, and $r > b$ and match the appropriate boundary conditions. However, for the purpose of calculating the tunneling probability it is sufficient to consider the solution inside the barrier. For $a < r < b$ we write the solution as the linear superposition of the positive and negative forms of (8.12) as

$$R(r) = \frac{A}{r}\exp\left(+\int_r^b F(r')dr'\right) + \frac{B}{r}\exp\left(-\int_r^b F(r')dr'\right). \tag{8.14}$$

The solution of equation (8.14) is dominated by the term with the coefficient A and this allows us to ignore the second term on the right in this equation.

The tunneling probability, P, that is the probability that an α-particle that has formed near the surface of the nucleus (that is, at radius a) will escape from the nucleus (that is, it will get to radius b), is given as the ratio of the probability that the particle will exist at b to the probability that it will exist at a. These probabilities are given as the square of the radial part of the wave function multiplied by the surface area. That is

$$P = \frac{4\pi b^2 |R(b)|^2}{4\pi a^2 |R(a)|^2} = e^{-G} \tag{8.15}$$

where the quantity of G is defined from the previous equation to be

$$G = 2 \int_a^b F(r') dr' = 2 \sqrt{\frac{2mQ}{\hbar^2}} \int_a^b \left(\frac{b}{r'} - 1 \right)^{1/2} dr'. \tag{8.16}$$

Integrating this expression yields

$$G = \frac{4Z_D e^2}{4\pi\varepsilon_0} \sqrt{\frac{2m}{\hbar^2 Q}} \left[\cos^{-1} \sqrt{\frac{a}{b}} - \sqrt{\frac{a}{b}\left(1 - \frac{a}{b}\right)} \right]. \tag{8.17}$$

This allows for the calculation of the barrier penetration probability in terms of known quantities.

It is now necessary to consider the time scale, τ_0, in equation (8.6). There are two factors that contribute to τ_0. The first depends on the details of the processes that cause the formation of the α-particle within the nucleus. Our knowledge of this is limited. The other factor may be viewed more or less classically in terms of an α-particle that is bouncing off the walls of the nuclear square well. The time scale for this process is related to the α-particle's velocity (which is related to the Q for the decay) and the radius of the nucleus. However, in order to consider the validity of the above description of α-decay we must make some assumptions concerning τ_0. We have no reason to expect that τ_0 will be substantially different in nuclei with similar mass and similar nucleon configurations. We will therefore consider the case where τ_0 is a constant and write

$$\tau = \frac{\tau_0}{e^{-G}}. \tag{8.18}$$

It is most appropriate to consider the α-decay of even–even nuclei. The results of applying the tunneling model to the α-decay of a number of even–even nuclei are given in Table 8.2. The calculated lifetimes have been normalized to

Table 8.2 | Measured and Calculated α-Decay Lifetimes for Some Heavy Nuclei

Parent	Daughter	Q (MeV)	τ_{meas} (s)	τ_{calc} (s)
^{238}U	^{234}Th	4.27	2.0×10^{17}	3.0×10^{17}
^{234}U	^{230}Th	4.86	1.1×10^{13}	1.0×10^{13}
^{230}Th	^{226}Ra	4.77	3.5×10^{12}	3.5×10^{12}
^{226}Ra	^{222}Rn	4.87	7.4×10^{10}	6.6×10^{10}
^{222}Rn	^{218}Po	5.59	4.8×10^5	3.8×10^5
^{218}Po	^{214}Pb	6.11	2.6×10^2	1.4×10^2
^{214}Po	^{210}Pb	7.84	2.3×10^{-4}	1.0×10^{-4}
^{210}Po	^{206}Pb	5.41	1.7×10^7	5.2×10^5

the lifetime of ^{230}Th and this gives the value of τ_0 from equation (8.18) as 6.3×10^{-23} s. With the exception of a slight discrepancy in the lifetime for ^{210}Po, the model results are amazingly consistent with experimentally measured values. The results presented in this section in the form of equations (8.17) and (8.18) provide a quantitative basis for the Geiger-Nutall rule. In cases where even–even nuclei decay by α-emission, the decay is almost always preferentially to the ground state of the daughter. This is reasonable, as the decay constant is such a sensitive function of the decay energy. If a nucleus were to decay to an excited state of the daughter, then the decay energy would be decreased proportionately, yielding a smaller branching ratio than a transition to the ground state. This is discussed further below.

The question of α-decay in even–odd, odd–even, and odd–odd nuclei is quite interesting. In general, lifetimes for these nuclei are as much as three orders of magnitude longer than those of even–even nuclei with similar mass and Q. The reason for this has to do with pairing effects and the formation of the α-particle within the nucleus. It appears that paired neutrons and protons, as exist in even–even nuclei, fairly readily form a bound α-particle state within the nucleus. However, unpaired nucleons do not readily participate in the formation of an α-particle. As a result the α-particle is most likely formed from lower lying paired nucleons leaving the unpaired nucleon in a higher energy state. This results in a daughter nucleus that is preferentially left in an excited state and thereby reduces the decay energy and increases the lifetime.

8.3 ANGULAR MOMENTUM CONSIDERATIONS

In the discussion above it has been assumed that $l = 0$. Certainly conservation laws impose some restrictions here and α-decay processes are allowed only if total angular momentum and parity are conserved. This means that the angular momenta and parities of the daughter nucleus and the α-particle must combine to yield the angular momentum and parity of the parent nucleus. Since the orientation of the angular momenta of the various nuclei involved in the decay are not specified, we can merely put limits on the orbital angular momentum of the α-particle, l, in terms of the values of J for the parent (P) and daughter (D) nuclei. That is, conservation of angular momentum requires

$$\left| \vec{J}_D - \vec{J}_P \right| \le l \le \left| \vec{J}_D + \vec{J}_P \right|. \tag{8.19}$$

As an example, we consider the α-decay ^{237}Np($^5/_2{}^+$) $\rightarrow$ ^{233}Pa($^3/_2{}^-$) $+ \alpha$. Conservation of angular momentum allows the α-particle to have $l = 1, 2, 3,$ or 4 although conservation of parity requires the α-particle to have odd parity, constraining l to be 1 or 3.

For nonzero l an inspection of equation (8.8) shows immediately that $l \ne 0$ has the effect of increasing the potential and constitutes a contribution to the overall potential barrier that the α-particle must tunnel through. This is the angular momentum barrier and it has the effect of reducing, in some cases substantially,

Table 8.3 | Relative Decay Rates for Different Values of α-Particle Angular Momentum

l	λ_l/λ_0
0	1.0
1	0.7
2	0.37
3	0.137
4	0.037
5	0.0071
6	0.0011

the value of the α-decay constant. An example of the decrease in the decay constant for α-decays producing α-particles with different l is given in Table 8.3.

The rotational levels of heavy nuclei are an ideal example of this kind of behavior. Figure 8.5 shows the branching ratios for the α-decay of ^{244}Cm to the ground state and the first few excited states of ^{240}Pu. Progressing from the ground state through the excited states of ^{240}Pu shows that the branching ratio for α-decay decreases consistently and substantially. This decrease is due to the

Figure 8.5 | Branching Ratios for the α-Decay of ^{244}Cm

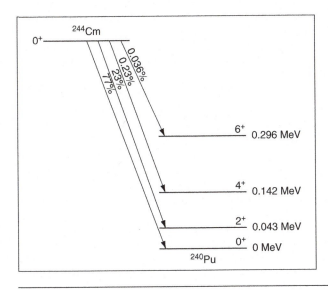

Figure 8.6 | Branching Ratios for the α-Decay of ^{243}Am

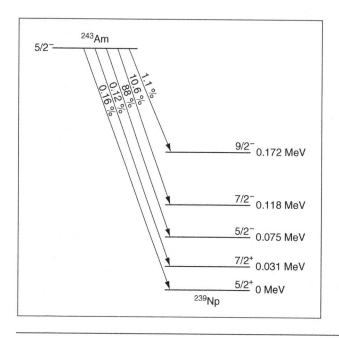

combined effects of an increase in the angular momentum barrier and a decrease in the α-particle energy. In even–even nuclei the transition to the ground state is preferred as the 0^+ to 0^+ transition eliminates the angular momentum term in the Schrödinger equation and the transition energy is maximized. In odd A nuclei the situation is sometimes different as shown in Figure 8.6 for the decay of ^{243}Am. Here the principal decay mode is to the 0.075 MeV state of the ^{239}Np daughter, since the elimination of the angular momentum barrier for the $(^5/_2{}^-)$ to $(^5/_2{}^-)$ transition more than compensates for the 0.075 MeV decrease in Q. Note that the ground state transition is not preferred because the change in parity does not allow the 0^+ state for the α-particle.

Problems

8.1. Tabulate all allowed α-particle spin and parity states for all transitions shown in Figures 8.5 and 8.6.

8.2. Using known atomic masses calculate the value of Q for the α-decay of the following nuclides. Assume transitions between ground states.
(a) ^{208}Po, (b) ^{222}Ra, (c) ^{240}Pu, and (d) ^{252}Fm.

8.3. From the *Table of Isotopes* (or another suitable reference) locate three examples of each of the following:

(i) the α-decay of an even–even nucleus

(ii) the α-decay of an even–odd nucleus

(iii) the α-decay of an odd–even nucleus

(iv) the α-decay of an odd–odd nucleus

Discuss the relationship between Q and lifetime for each of these cases.

8.4. The lifetime for the α-decay of ^{226}Ra is 7.3×10^{10} s. Use this information to calculate the radius of a ^{222}Rn nucleus.

8.5. Calculate the kinetic energy of the α-particle released during the α-decay of ^{225}Ac.

8.6. (a) Consider the α-decay of a highly nonspherical nucleus (that is, one that has a large electric quadrupole moment). Calculate the relative α-decay rates along the semimajor axis and semiminor axis of a nucleus with a semimajor to semiminor radius ratio of 1.5.

(b) Describe the α-decay radiation pattern that is expected from such a nucleus.

8.7. The α-decay of even–even parent nuclides preferentially populates the ground state of the daughter (rather than an excited state of the daughter). This is not necessarily the case for even–odd or odd–even parents. Explain.

8.8. (a) Calculate the Q for the α-decay of ^{241}Cm.

(b) The branching ratios for the α-decay of ^{241}Cm to the excited states of ^{237}Pu are 13% to the 0.202 MeV state ($^5/_2{}^+$), 17% to the 0.156 MeV state ($^3/_2{}^+$), and 70% to the 0.145 MeV state ($^1/_2{}^+$). There is no decay to the ^{237}Pu ground state. Explain.

8.9. Using the semiempirical mass formula, derive an expression for the Q of an α-decay process. Use this expression to calculate the Q for the α-decay of ^{241}Am and compare with the value based on measured atomic masses.

8.10. (a) The ground state of ^{8}Be decays by splitting into two α-particles. Calculate the Q for this process.

(b) Excited states of ^{12}C are known to decay by splitting into three α-particles. Calculate the threshold energy for this process.

(c) Consider the possibility of the decay of excited states of ^{16}O into four α-particles.

9 CHAPTER | **Beta Decay**

9.1 ENERGETICS OF BETA DECAY

Beta decay (β-decay) as described in Chapter 4 represents the conversion of a neutron to a proton or the conversion of a proton to a neutron. The former situation is referred to as negative β-decay (β^--decay) and from a nuclear standpoint is described as

$$\,_Z^A X^N \rightarrow \,_{Z+1}^A Y^{N-1} + e^- + \bar{\nu}_e. \tag{9.1}$$

The conversion of a proton to a neutron represents positive β-decay (β^+-decay) and can be written as a nuclear process as

$$\,_Z^A X^N \rightarrow \,_{Z-1}^A Y^{N+1} + e^+ + \nu_e. \tag{9.2}$$

In both cases the appearance of the neutrino (or antineutrino) is necessary in order to conserve lepton number. Its existence is also needed to properly explain the distribution of electron (or positron) energies as discussed below. Another process which, in many ways, is analogous to β^+-decay is electron capture. This corresponds to the fundamental process

$$p + e^- \rightarrow n + \nu_e. \tag{9.3}$$

In this case a proton in the nucleus interacts with an atomic electron (typically an inner shell s-electron because the wave function will have the greatest probability of overlapping with the nuclear wave function). This represents the nuclear process

$$_{Z}^{A}X^{N} + e^{-} \rightarrow _{Z-1}^{A}Y^{N+1} + v_{e}$$ (9.4)

where it is seen that the identity of the parent and daughter nuclei are the same as for β^{+}-decay.

Here we consider the energetics of these three processes. For β^{-}-decay we write the energy released, Q, in terms of the nuclear masses (subscript N):

$$Q = \left[m_{N}\left(_{Z}^{A}X^{N}\right) - m_{N}\left(_{Z+1}^{A}Y^{N-1}\right) - m_{e} \right]c^{2}.$$ (9.5)

It is assumed that the neutrino mass is negligible and that the transition is between nuclear ground states. The validity of this former assumption is discussed further in Chapter 17. If we ignore the electronic binding energy then equation (9.5) can be written in terms of atomic masses using the relation

$$m\left(_{Z}^{A}X^{N}\right) = m_{N}\left(_{Z}^{A}X^{N}\right) + Zm_{e}.$$ (9.6)

This gives

$$Q = \left[m\left(_{Z}^{A}X^{N}\right) - m\left(_{Z+1}^{A}Y^{N-1}\right) \right]c^{2}.$$ (9.7)

The omission of the electronic binding energy is justified since it is only the difference in binding energies between the right and left sides of equation (9.1) that is important and this will be negligibly small.

For β^{+}-decay the Q can be written in terms of nuclear masses from equation (9.2) as

$$Q = \left[m_{N}\left(_{Z}^{A}X^{N}\right) - m_{N}\left(_{Z-1}^{A}Y^{N+1}\right) - m_{e} \right]c^{2},$$ (9.8)

noting that the positron and the electron have the same mass. Converting this expression to atomic masses gives

$$Q = \left[m\left(_{Z}^{A}X^{N}\right) - m\left(_{Z-1}^{A}Y^{N+1}\right) - 2m_{e} \right]c^{2}.$$ (9.9)

The energetics of electron capture is given by equation (9.4) as

$$Q = \left[m_{N}\left(_{Z}^{A}X^{N}\right) + m_{e} - m_{N}\left(_{Z-1}^{A}Y^{N+1}\right) \right]c^{2} - b_{e}$$ (9.10)

where b_{e} is the binding energy of the electron that is captured. In addition to the electronic binding energy terms that more or less cancel out in equations such as (9.7), equation (9.10) contains a term from the binding energy of the electron that is captured. As this is typically an inner shell (K-shell) electron then this energy can be significant (many tens of keV) and is often included in the energy calculation for electron capture. In terms of atomic masses the above becomes

$$Q = \left[m\left(_{Z}^{A}X^{N}\right) - m\left(_{Z-1}^{A}Y^{N+1}\right) \right]c^{2} - b_{e}.$$ (9.11)

Figure 9.1 | Energy Spectrum of Electrons from the β^--Decay of ^{64}Cu

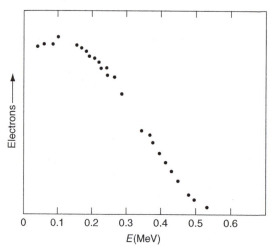

Data have been calculated from results given in L. M. Langer, R. D. Moffat, and H. C. Price, Jr., *Phys. Rev.* 76 (1949), 1725.

When Q is positive, the process is exothermic and can proceed spontaneously. When Q is negative, the process is endothermic and is not energetically favorable. A simple inspection of equations (9.9) and (9.11) shows that electron capture is energetically more favorable than β^+-decay by an additional $2m_e c^2$. An investigation of the nuclear data tables shows that for some β-unstable nuclei electron capture occurs (because Q in equation (9.11) is positive) while β^+-decay does not occur (because Q in equation (9.9) is negative).

In a process such as that shown in equation (9.1), the energy that is given up during the decay process appears as the kinetic energy of the electron and antineutrino and the recoil energy of the daughter nucleus. Because of the large mass associated with the daughter nucleus its recoil energy is substantially smaller than the precision with which the electron's energy can be measured. Thus in subsequent discussions in this chapter the nuclear recoil energy will be ignored. Thus for the process in (9.1), we consider the energy as being distributed between the electron and the antineutrino. ^{64}Cu is an interesting example as it decays by both β^--decay to ^{64}Zn and by β^+-decay or electron capture to ^{64}Ni. The energy spectrum of electrons emitted by the β^--decay of ^{64}Cu is illustrated in Figure 9.1. The plot shows a distribution of electron energies up to slightly less than 0.6 MeV. This maximum energy is referred to as the endpoint energy and for ground state to ground state transitions is the value of Q as obtained from equation (9.7). A simple calculation based on tabulated atomic masses gives the endpoint energy of 0.579 MeV, in agreement with the figure.

Figure 9.2 | Energy Spectrum of Positrons from the β^+-Decay of ^{64}Cu

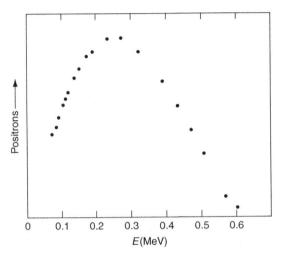

Data have been calculated from results given in L. M. Langer, R. D. Moffat, and H. C. Price, Jr., *Phys. Rev.* 76 (1949), 1725.

Decays that produce an electron with the endpoint energy also produce an antineutrino with negligibly small energy. On the other hand decays that emit an electron with small energy emit an antineutrino with close to the endpoint energy. The measurement shown here indicates that it is much more probable for the electron to have a small amount of energy and the antineutrino to get the majority than the other way around.

The measured energy spectrum of the positrons emitted from the β^+-decay of ^{64}Cu is shown in Figure 9.2. A simple calculation using measured atomic masses in equation (9.9) gives a positron endpoint energy of 0.653 MeV, consistent with the data shown in the figure.

The theoretical model of β-decay as described below has been developed to explain the form of the energy spectrum as shown in Figures 9.1 and 9.2 and to explain the observed lifetimes for β-decay processes.

9.2 FERMI THEORY OF BETA DECAY

The decay constant for a decay process is merely the transition rate from the initial (parent) to the final (daughter) state. Perturbation theory gives the transition rate between states (Fermi's Golden Rule) as

$$\lambda = \frac{2\pi}{\hbar} \left| H_{if} \right|^2 \frac{dn}{dE} \qquad (9.12)$$

where dn/dE is the density of final states and H_{if} is a matrix element given by the integral of the interaction operator, H:

$$H_{if} = G \int \psi_f^* H \psi_i \, d^3 \vec{r}. \qquad (9.13)$$

Here ψ_i and ψ_f are the wave functions of the initial and final states, respectively, and G is a constant representing the strength of the interaction. We will consider β^--decay as described in equation (9.1). The wave function of the initial state is the wave function of the parent nuclear state (P). The wave function of the final state is given by the wave functions of the daughter nuclear state (D), the electron, and the antineutrino so that equation (9.13) can be written as

$$H_{if} = G \int \psi_D^* \psi_e^* \psi_{\bar{v}_e}^* H \psi_P \, d^3 \vec{r}. \qquad (9.14)$$

A similar matrix element could also be constructed for β^+-decay. The electron and antineutrino wave functions are taken to be plane waves (because these are free particles) and, as a matter of convenience, are normalized over the nuclear volume, V:

$$\psi_e(\vec{r}) = \frac{1}{\sqrt{V}} e^{i\vec{p}\cdot\vec{r}/\hbar} \qquad (9.15)$$

and

$$\psi_{\bar{v}_e}(\vec{r}) = \frac{1}{\sqrt{V}} e^{i\vec{q}\cdot\vec{r}/\hbar}. \qquad (9.16)$$

Here the p and q are the momentum of the electron and the antineutrino, respectively. The exponential in equation (9.15) can be expanded as

$$e^{i\vec{p}\cdot\vec{r}/\hbar} = 1 + \frac{i\vec{p}\cdot\vec{r}}{\hbar} + \cdots \qquad (9.17)$$

and similarly for the antineutrino. For electrons with typical β^--decay energies, p is of the order of 1 MeV/c and for r of the order of the nuclear radius we find $pr/\hbar = 0.03$. Thus, taking only the leading terms in the expansions for the wave functions in equations (9.15) and (9.16) would seem to be a reasonable approximation and gives

$$H_{if} = \frac{GM_{if}}{V} \qquad (9.18)$$

where we have defined

$$M_{if} = \int \psi_D^* H \psi_P \, d^3 \vec{r}. \qquad (9.19)$$

It is now necessary to consider the question of the density of final states. We can consider a simple analysis based on a free-particle Schrödinger equation:

$$-\frac{\hbar^2}{2m}\nabla^2\psi = E\psi. \tag{9.20}$$

For the simple case of a three-dimensional infinite well in Cartesian coordinates this equation has momentum states that are quantized as

$$p_i = \frac{n_i\pi\hbar}{L} \tag{9.21}$$

where i is a coordinate, x, y, or z, and L is the length of the edge of the well. This means that the distance between allowed states in momentum space is $\pi\hbar/L$ and the density of states is therefore $L/\pi\hbar$ along each Cartesian direction. The density in three dimensions will be given by the cube of this quantity. The total number of allowed states with momentum less than a value $p = (p_x^2 + p_y^2 + p_z^2)^{1/2}$ is given by this density of states and the volume of a sphere in momentum space as

$$n_e = \left(\frac{1}{8}\right)\left(\frac{4}{3}\pi p^3\right)\left(\frac{L}{\pi\hbar}\right)^3. \tag{9.22}$$

The additional factor of $1/8$ comes from the fact that only one quadrant in p space needs to be counted, as changing the sign of n_i in equation (9.21) does not yield new independent states. For the electron the density of states per unit momentum is found by differentiating (9.22) to give

$$dn_e = \frac{4\pi V p^2 dp}{(2\pi\hbar)^3}. \tag{9.23}$$

Correspondingly the density of states for the antineutrino is

$$dn_{\bar{\nu}_e} = \frac{4\pi V q^2 dq}{(2\pi\hbar)^3}. \tag{9.24}$$

In order to determine the transition rate as given by equation (9.12) it is necessary to consider the case where the final state is characterized by the correct electron and antineutrino energies and momenta. Specifically we can write the antineutrino density of states corresponding to a particular final state energy as $dn_{\bar{\nu}_e}/dE_f$. The partial transition rate into the correct electron momentum state is therefore

$$d\lambda = \frac{2\pi}{\hbar}G^2|H_{if}|^2\frac{dn_{\bar{\nu}_e}}{dE_f}dn_e. \tag{9.25}$$

In principle the total transition rate can be obtained by integrating the above, but this is not necessary in order to obtain a result that can be compared with the experimental energy spectrum. Substituting (9.18), (9.23), and (9.24) into (9.25) gives

$$d\lambda = \frac{G^2}{2\pi^3\hbar^7}\left|M_{if}\right|^2 p^2 q^2 \frac{dq}{dE_f}\,dp. \tag{9.26}$$

It is important to observe that the nuclear volume that appeared in the above calculations has cancelled out. The dq/dE_f term on the right-hand side above can be dealt with by writing the total final state energy as

$$E_f = Q + m_e c^2 \tag{9.27}$$

where the Q of the decay is written as the sum of the kinetic energies of the electron and the antineutrino:

$$Q = T_e + qc. \tag{9.28}$$

Using equation (9.28) in (9.27) and keeping T_e constant yields

$$\frac{dq}{dE_f} = \frac{1}{c} \tag{9.29}$$

and

$$q = \frac{Q - T_e}{c}. \tag{9.30}$$

Substituting equations (9.29) and (9.30) into (9.26) yields

$$\frac{d\lambda}{dp} = \frac{G^2}{2\pi^3\hbar^7 c^3}\left|M_{if}\right|^2 p^2 (Q - T_e)^2. \tag{9.31}$$

Experimentally, as shown in Figure 9.1, the spectral intensity is measured as a function of energy, in this case the kinetic energy of the electron. Equation (9.31) can be rewritten in terms of energy. In general, the Q associated with β-decay electron is comparable to the electron rest mass energy and it is necessary to include relativistic effects in this calculation. Thus the electron kinetic energy is expressed as

$$T_e = \sqrt{p^2 c^2 + m_e^2 c^4} - m_e c^2 \tag{9.32}$$

Figure 9.3 | Calculated Electron Energy Spectrum for β-Decay

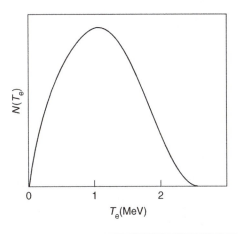

and the measured spectral intensity is given as the energy dependent decay constant:

$$I(E) = \frac{d\lambda}{dT_e} = \frac{G^2}{2\pi^3 \hbar^7 c^6} \left| M_{if} \right|^2 (T_e^2 + 2T_e m_e c^2)^{1/2} (Q - T_e)^2 (T_e + m_e c^2). \quad (9.33)$$

In order to make some numerical predictions using this theory it is necessary to have some information concerning the dependence of M_{if} on the electron kinetic energy (or equivalently on p and q). The simplest assumption that we can make is that M_{if} is independent of T_e and it turns out that this is applicable to a large number of β-decay processes. The electron energy spectrum as calculated from equation (9.31) using this approximation is shown in Figure 9.3. Thus far we have included no factors that distinguish the electron from the positron so we expect the same result as shown in the figure for the positron energy spectrum from β^+-decay. Figure 9.3 shows somewhat better agreement with the positron spectrum in Figure 9.2 than with the electron spectrum in Figure 9.1. However, careful inspection shows that there is some discrepancy at low energy in both cases. There is a simple explanation for this behavior. Once produced by the decay of a proton the positively charged positron is repelled by the coulombic interaction with the daughter nucleus. Recall that the positron, being a lepton, is not influenced by the strong nuclear force. This repulsive interaction literally pushes the positron out of the nucleus and diminishes the low energy portion of the spectrum. The electron, on the other hand, after being created by the decay of a neutron, is held back by the attractive coulombic force with the daughter nucleus. This, therefore, augments the low energy portion of the spectrum. These coulombic effects can be taken into account in a more quantitative way by introducing an additional factor, $F(Z, T_e)$, in equation (9.33) as

$$I(E) = \frac{G^2}{2\pi^3\hbar^7 c^6}\left|M_{if}\right|^2\left(T_e^2 + 2T_e m_e c^2\right)^{1/2}\left(Q - T_e\right)^2\left(T_e + m_e c^2\right)F(Z_D, T_e). \quad (9.34)$$

$F(Z, T_e)$ is called the Coulomb factor or Fermi factor (or function) and is a function of the electron (or positron) kinetic energy and the charge on the daughter nucleus. A nonrelativistic calculation yields

$$F(Z_D, T_e) = \frac{2\pi\eta}{1 - \exp(-2\pi\eta)} \quad (9.35)$$

where

$$\eta = \pm\frac{Z_D e^2}{4\pi\varepsilon_0 \hbar v}. \quad (9.36)$$

The electron/positron velocity is given by v and the positive sign is used in this expression for electrons (β^--decay) and the negative sign for positrons (β^+-decay). This expression is appropriate at low energies where the differences between Figures 9.1 and 9.2 and Figure 9.3 are most apparent. Results of a more general calculation that includes relativistic effects can be found in *Tables for the Analysis of Beta Spectra*, National Bureau of Standards, Applied Mathematical Series, 13. Inclusion of this factor clearly shows that the low energy portion of the electron spectrum for β^--decay will be augmented while this portion of the positron spectrum for β^+-decay will be diminished.

9.3 FERMI-KURIE PLOTS

The validity of this theory may be tested using a Fermi-Kurie plot. This is sometimes also called a Kurie plot or Fermi plot. The spectral intensity as a function of electron momentum is given from equation (9.31) combined with the Coulomb factor as

$$I(p) = \frac{G^2}{2\pi^3\hbar^7 c^3}\left|M_{if}\right|^2 p^2\left(Q - T_e\right)^2 F(Z_D, T_e). \quad (9.37)$$

Plotting $[I(p)/p^2 F]^{1/2}$ as a function of T_e will yield a straight line if the above assumptions of the theory are correct. The intercept on the horizontal axis will immediately give the value of Q. In cases where the decay is to an excited state of the daughter nucleus then it is essential to properly include this in a consideration of the electron energies. An example of a Fermi-Kurie plot for the

Figure 9.4 | Fermi-Kurie Plot for the β^--Decay of ^{3}H

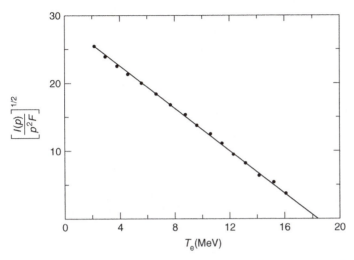

From S. C. Curran, J. Angus, and A. L. Cockroft, *Phil. Mag. 40* (1949), 53. Copyright Taylor & Francis, Ltd. Used with permission (http://www.tandf.co.uk/journals).

β^--decay of ^{3}H (tritium) is shown in Figure 9.4. The linear relationship shown in this plot is evidence that our interpretation of the energy dependence of the right-hand side of equation (9.34) was correct. This energy dependence was based on our elimination of higher order terms in the expansion of the electron and antineutrino wave functions in equation (9.17). Beta transitions that are properly described by the linear relationship shown in the figure are said to be *allowed transitions*. This terminology is somewhat ambiguous as transitions that do not follow this linear behavior can still occur. This is discussed further in Section 9.4. An example of a decay that is not described by the Fermi-Kurie plot is illustrated in Figure 9.5a. This decay is referred to as *first-forbidden* (although it is obviously not forbidden entirely). The nonlinearities that are clearly seen are the result of additional energy dependence coming into equation (9.34). This nonlinearity can be accounted for by the inclusion of an additional term, $S(p, q)$, sometimes called a shape factor, in equation (9.37). In the case of the first forbidden decay shown in Figure 9.5a, the shape factor can be expressed as

$$S(p, q) \propto p^2 + q^2 \qquad (9.38)$$

and comes from the next highest order terms in expansions such as equation (9.17). Including this in equation (9.37) and plotting $[I(p)/p^2FS]^{1/2}$ as a function of energy as illustrated in Figure 9.5b shows the expected linear dependence.

Figure 9.5 | (a) Fermi-Kurie Plot for the β^--Decay of ^{91}Y and
(b) Corrected Fermi-Kurie Plot for the β^--Decay of ^{91}Y
Using the Shape Factor for a First Forbidden Decay

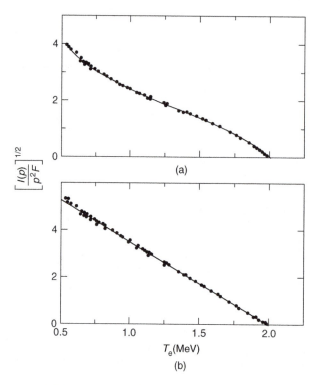

From L. M. Langer and H. C. Price, Jr., *Phys. Rev. 75* (1949), 1109. Copyright 1949 by the American Physical Society.

9.4 ALLOWED AND FORBIDDEN TRANSITIONS

The question of allowed and forbidden transitions can be described in more detail by considering angular momentum conservation. The elimination of higher order terms from equation (9.17) is valid in the limit as $r \to 0$. In this limit the electron (and antineutrino) can be viewed as being created at $r = 0$ and can therefore carry no orbital angular momentum. The total angular momentum of the electron-antineutrino pair is, therefore, the vector sum of their individual spins. Because the intrinsic spin is $s = 1/2$ then the total spin can be either $S = 0$ or $S = 1$. The spin 0 and spin 1 cases are referred to as Fermi decays and Gamow-Teller decays and correspond to singlet and triplet states for the electron-antineutrino, respectively. Conservation of total angular momentum requires that the change in J between the parent and daughter be related to the spin of the

Table 9.1 | Properties of Allowed and Forbidden β-Decays

Decay	L	ΔJ	Nuclear Parity Change
allowed	0	0, ±1	no
1st forbidden	1	0, ±1, ±2	yes
2nd forbidden	2	±1, ±2, ±3	no
3rd forbidden	3	±2, ±3, ±4	yes
4th forbidden	4	±3, ±4, ±5	no

electron-antineutrino pair as

$$\left| \vec{J}_{\mathrm{D}} - \vec{J}_{\mathrm{P}} \right| \le S \le \left| \vec{J}_{\mathrm{D}} + \vec{J}_{\mathrm{P}} \right|. \tag{9.39}$$

For allowed decays the above shows that ΔJ defined as $J_{\mathrm{D}} - J_{\mathrm{P}}$ can be 0 or ±1. Fermi and Gamow-Teller transitions can occur for all cases that can satisfy equation (9.39) as indicated in Table 9.1. However, it is important to note that only Fermi transitions are allowed for the case $J_{\mathrm{D}} = J_{\mathrm{P}} = 0$. Conservation of parity requires no change in parity between the parent and daughter nuclei since the parity of the electron-antineutrino pair (with $L = 0$) must be even.

Forbidden decays are not really forbidden but are merely less probable than allowed decays and therefore correspond to longer lifetimes. If the matrix element in equation (9.12) that corresponds to the allowed transition vanishes, then the only transition that can occur is a forbidden decay. This requires the inclusion of higher order terms in the expansion of equation (9.17). An example of the first forbidden decay where the allowed matrix element vanishes but the next order term in the expansion does not vanish was seen in the previous section. If the first two matrix elements vanish then the next order term must be included and this is referred to as a second forbidden decay. This terminology continues to higher order forbidden decays. In all cases the lowest order nonvanishing term dominates and can be accounted for in the calculation of the energy spectrum by inclusion of the appropriate shape factor. Shape factors for the first few forbidden decays as obtained from the expansion of the exponential in equations (9.15) and (9.16) are given in Table 9.2.

Table 9.2 | Shape Factors for Forbidden β-Decays

Decay	$S(p, q)$
1st forbidden	$(m_e c)^{-2}(p^2 + q^2)$
2nd forbidden	$(m_e c)^{-4}[p^4 + q^4 + (10/3)p^2 q^2]$
3rd forbidden	$(m_e c)^{-6}[p^6 + q^6 + 7p^2 q^2(p^2 + q^2)]$

The most obvious cases where a first forbidden transition can occur is a situation where there is a change of parity between the parent and daughter nuclei. In this case the electron-antineutrino must have odd parity and since $\pi = (-1)^L$, L cannot be zero. This requires that r cannot be approximated as zero in equation (9.17) and higher order terms must be included. Possible values for ΔJ consistent with equation (9.39) for first forbidden decays are given in Table 9.1.

In cases where $\Delta J = \pm 2$ and there is no parity change or $\Delta J \geq 3$ then an allowed or first forbidden decay cannot satisfy the necessary conservation laws. In this case a second or higher order forbidden decay may occur. Some of these are summarized in Table 9.1. In all cases forbidden decays of a given order are less probable than those of the previous order; typically by a factor of about 10^3 for decays with similar Q.

The distinction between allowed and forbidden decays can also be evaluated on the basis of equation (9.37). This is the decay rate per unit momentum so the total decay rate can be determined by integrating the right-hand side of this expression over momentum. This gives

$$\lambda = \int_0^{p_{max}} I(p)\,dp = \frac{m_e^5 c^4 G^2}{2\pi^3 \hbar^7}\left|M_{if}\right|^2 f(Z_D, Q) \tag{9.40}$$

where $f(Z_D, Q)$ is defined as

$$f(Z_D, Q) = \int_0^{p_{max}} F(Z_D, T_e)\frac{p^2}{m_e^2 c^2}\frac{(Q - T_e)^2}{m_e^2 c^4}\frac{dp}{m_e c}. \tag{9.41}$$

The additional factors of m_e and c have been included in order to make the integrand dimensionless. The integration is up to a maximum value of momentum given (relativistically) in terms of Q as

$$p_{max} = \frac{1}{c}\left[(Q + m_e c^2)^2 - m_e^2 c^4\right]^{1/2}. \tag{9.42}$$

Numerically determined values of the function $f(Z_D, Q)$ for some typical values of Z_D and Q are shown in Figure 9.6. From the relationship between the decay rate and the halflife, equation (7.3), the above gives

$$f(Z_D, Q)\tau_{1/2} = \frac{2\pi^3 \hbar^7 \ln 2}{m_e^5 c^4 G^2}\frac{1}{\left|M_{if}\right|^2}. \tag{9.43}$$

This expression shows that the value of $f(Z_D, Q)\tau_{1/2}$ depends only on the matrix element M_{if}. It is generally most convenient to deal with $\log[f(Z_D, Q)\tau_{1/2}]$ and we would expect that allowed transitions would have similar values for this quantity because of similar matrix elements for the transitions. An analysis of $\log[f(Z_D, Q)\tau_{1/2}]$ for different decays provides some useful information about the degree to which the transition is forbidden. Figure 9.7 shows $\log[f(Z_D, Q)\tau_{1/2}]$

Figure 9.6

The Logarithm of [f(Z_D, Q)], as Given by Equation (9.41), as a Function of Q for Some Typical Values of Z_D

Values are shown for both β^+- and β^--decay.

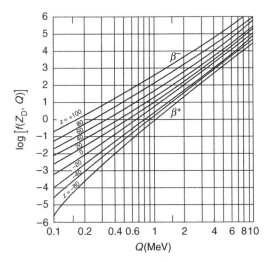

From R. D. Evans, *The Atomic Nucleus* (Melbourne, FL: Krieger Publishing, 1982).

Figure 9.7

The Quantity log [f(Z_D, Q)τ_{1/2}] (with $\tau_{1/2}$ in Seconds) Plotted as a Function of the Degree of Forbiddenness for Some Known β^--Decays

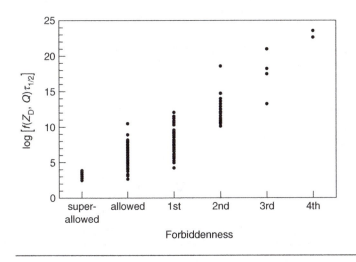

Table 9.3 | Typical Ranges of Log $[f(Z_D, Q)\tau_{1/2}]$ Values for Different β-Decays

Decay	$\log [f(Z_D, Q)\tau_{1/2}]$ $[\tau_{1/2}$ in seconds$]$
superallowed	3–4
allowed	4–8
1st forbidden	6–9
2nd forbidden	10–14
3rd forbidden	15–21
4th forbidden	21–24

for some known transitions. It would be convenient to find well-defined and different $\log [f(Z_D, Q)\tau_{1/2}]$ values for decays of different forbiddenness. Unfortunately the distinction is not always clear as seen in Figure 9.7. In terms of increasing $\log [f(Z_D, Q)\tau_{1/2}]$ values the transitions with the smallest values are sometimes referred to as *superallowed*. These are followed by allowed transitions and transitions of increasing degree of forbiddenness as shown in Table 9.3.

This analysis has some practical applications for the understanding of transitions between nuclear states. It is convenient to write

$$\log[f(Z_D, Q)\tau_{1/2}] = \log[f(Z_D, Q)] + \log[\tau_{1/2}]. \tag{9.44}$$

Experimentally it is straightforward to determine $\tau_{1/2}$ and the decay energy Q. From Figure 9.6 $\log [f(Z_D, Q)]$ can be determined and this allows $\log [f(Z_D, Q)\tau_{1/2}]$ to be calculated. Based on this value, the information in Table 9.3 permits us to make some educated guesses about the degree of forbiddenness. This is helpful when studying unknown transitions as it allows us to make some guesses concerning spin states. As an example of the calculation of $\log [f(Z_D, Q)\tau_{1/2}]$ let's consider the ground state to ground state β^--decay of ^{10}Be $\rightarrow$ ^{10}B + e$^-$ + $\bar{\nu}_e$. This has a measured halflife of 2.5×10^6 y $= 7.85 \times 10^{13}$ s. The decay energy is measured to be $Q = 0.555$ MeV and Figure 9.6 gives $\log [f(Z_D, Q)] = -0.2$. This yields a calculated value of $\log [f(Z_D, Q)\tau_{1/2}] = 13.7$. An analysis of the spin states shows this to be a second forbidden $0^+ \rightarrow 3^+$ transition, consistent with the observed $\log [f(Z_D, Q)\tau_{1/2}]$ value.

9.5 PARITY VIOLATION IN BETA DECAY

Beta decay results from the weak interaction and as indicated in Chapter 2 is a process that does not necessarily conserve parity. If a process conserves parity then the parity transform of any real experiment will yield the same results as the experiment itself. One way of observing these features is to investigate the

Figure 9.8 | Comparison of the Emission of Electrons from the β^--Decay of ^{60}Co as Seen in a Real Experiment (left) and in a Mirror (right)

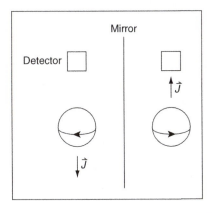

properties of the mirror image of an experiment. The possibility of observing these effects in processes that are governed by weak interactions was first proposed in 1956 by T.D. Lee and C.N. Yang. The first experimental evidence demonstrating nonconservation of parity in a weak process, namely β-decay of ^{60}Co, was reported in 1958 by C.S. Wu and coworkers. As described below the details of the experiment depend on the magnetic properties of cobalt.

Cobalt atoms possess a large atomic magnetic moment that can be readily aligned in an applied magnetic field. This atomic moment will couple to the nuclear magnetic moment and cause at least a partial alignment of the nuclear moments. This alignment is increased as the temperature is decreased due to a reduction in the thermal motion of the moments and, at sufficiently low temperatures (typically 0.01 K), a substantial fraction of the nuclear moments can be aligned. A measurement of the angular distribution of electrons from the β-decay of magnetically aligned ^{60}Co can be related to the preferential orientation of the emitted electrons and the direction of the nuclear magnetic moment. The case where there is a preferential emission antiparallel to the direction of the nuclear magnetic moment in the real experiment is illustrated in Figure 9.8. It turns out that this is what is observed experimentally. The experiment as viewed in the mirror is illustrated on the right-hand side of Figure 9.8 and it is seen in this case that the direction of the nuclear moment is opposite to that in the real experiment. Thus the real experiment and the experiment as seen in the mirror do not yield the same results. This means that the assumed anisotropy of the electron emission violates parity conservation. From an experimental standpoint, the mirror image of the real experiment is created by changing the direction of the applied magnetic field by 180°. Results of such an experiment are shown in Figure 9.9. The experiment was conducted by cooling the sample in an external field in order to align the nuclear moments. The sample was then

Figure 9.9 | Results of Parity Violation Experiments for the β^--Decay of ^{60}Co

Measurements were made for two magnetic field directions, indicated by the open and closed squares. Data were obtained as a function of time as the sample warmed; thus the horizontal axis represents increasing temperature from left to right.

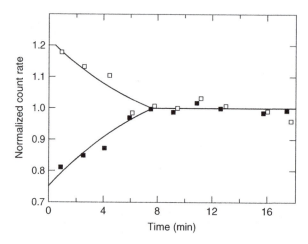

From C. S. Wu *et al.*, *Phys. Rev.* **105** (1958), 1413. Copyright 1958 by the American Physical Society.

allowed to warm up and the resulting thermal fluctuations randomized the ordering of the nuclear spins. The experiment was then repeated with the magnetic field applied in the opposite direction. The differences in the electron fluxes between the two field orientations is a clear demonstration of parity nonconservation. The nonconservation of parity in weak processes will be discussed further in Chapter 15 as it applies to the decay of mesons.

9.6 DOUBLE BETA DECAY

The possibility of an interesting decay process arises in situations such as that shown in Figure 4.5. Traditionally it is believed that ^{128}Te is β-stable since it is energetically unfavorable for it to decay either to ^{128}I by β^--decay or to ^{128}Sb by β^+-decay or electron capture. However, ^{128}Xe has a smaller mass than ^{128}Te and it would be energetically favorable for two ^{128}Te neutrons to become two protons and emit two electrons (and two antineutrinos). This double β-decay process is

$$^{128}\text{Te} \rightarrow {}^{128}\text{Xe} + 2e^- + 2\bar{\nu}_e \qquad (9.45)$$

and corresponds to the basic process

$$2n \rightarrow 2p + 2e^- + 2\bar{\nu}_e. \qquad (9.46)$$

Table 9.4 | Possible Double β-Decays

Parent	Parent Natural Abundance (%)	Daughter	Q (MeV)	Barrier Energy (MeV)
^{46}Ca	0.003	^{46}Ti	0.985	1.382
^{70}Zn	0.62	^{70}Ge	1.01	0.653
^{76}Ge	7.67	^{76}Se	2.041	0.92
^{80}Se	49.82	^{80}Kr	0.138	1.87
^{82}Se	9.19	^{82}Kr	3.00	0.09
^{86}Kr	17.37	^{86}Sr	1.24	0.054
^{94}Zr	2.8	^{94}Mo	1.23	0.921
^{100}Mo	9.62	^{100}Ru	3.03	0.335
^{104}Ru	18.5	^{104}Pd	1.32	1.15
^{110}Pd	12.5	^{110}Cd	2.0	0.87
^{114}Cd	28.86	^{114}Sn	0.547	1.44
^{116}Cd	7.58	^{116}Sn	2.81	0.52
^{122}Sn	4.71	^{122}Te	0.349	1.622
^{124}Sn	5.98	^{124}Te	2.26	0.653
^{128}Te	31.79	^{128}Xe	0.876	1.26
^{130}Te	34.49	^{130}Xe	2.54	0.41
^{134}Xe	10.44	^{134}Ba	0.73	1.33
^{136}Xe	8.87	^{136}Ba	2.72	0.112
^{142}Ce	11.07	^{142}Nd	1.379	0.777
^{148}Nd	5.7	^{148}Sm	1.936	0.514
^{150}Nd	5.6	^{150}Sm	3.39	0.036
^{154}Sm	22.6	^{154}Gd	1.26	0.72
^{160}Gd	21.75	^{160}Dy	1.78	0.029
^{238}U	99.28	^{238}Pu	1.173	0.117

The theoretically predicted lifetime is sufficiently long that observation of this decay will, at best, be very difficult. There are a number of possible candidates for observing double β-decay. As in the case of ^{128}Te, double β-decay is always from an even–even parent nucleus to an even–even daughter nucleus. The ground state transitions are, therefore, all $0^+ \rightarrow 0^+$. Possible double β-decay candidates are summarized in Table 9.4. The decay rate is expected to increase with increasing decay energy and also to increase with decreasing barrier height; that is the energy (mass) barrier formed by the odd–odd nucleus (^{128}I in Figure 4.5) between the parent and daughter nuclei.

Table 9.5 | Double β-Decay Lifetimes Determined by Geological Experiments and Calculated Decay Rates

Decay	Lifetime (years)	Decays per Year per Gram
^{130}Te $\rightarrow$ ^{130}Xe	3.2×10^{21}	0.503
^{128}Te $\rightarrow$ ^{128}Xe	5.0×10^{24}	0.00029
^{82}Se $\rightarrow$ ^{82}Kr	3.7×10^{20}	1.8

The first experimental evidence for double β-decay was based on geological observations of rock samples containing a possible double β-decay parent nuclide. In such samples it is expected that an excess of the daughter nuclide (relative to the other isotopes of that element) would be found. The details of the analysis of this kind of experiment follow along the lines of our discussion of radioactive dating in Section 7.3, although here we assume that we know the age of the rock and solve for the lifetime of the decay process. Some results of double β-decay lifetimes based on geological observations are given in Table 9.5.

A number of experiments designed to directly observe the energy spectrum of electrons produced by a double β-decay process have been conducted and many are still in progress. The very long lifetimes mean very low decay rates and subsequently experiments are difficult because of interference from background radiation. Many experiments are conducted underground in mines to shield them from cosmic ray interference. ^{82}Se is particularly interesting as the decay energy is fairly large and the barrier energy is very small. The energy level diagram for this decay is shown in Figure 9.10. Results of an experiment on ^{82}Se

Figure 9.10 | Possible Double β-Decay of ^{82}Se

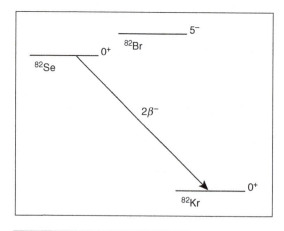

Figure 9.11 | Measured (data points) and Calculated (solid line) Electron Energy Spectrum from the Double β-Decay of ^{82}Se

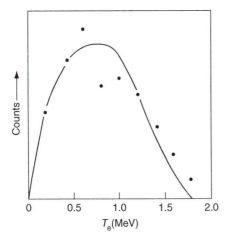

Data are from S. R. Elliot, A. A. Hahn, and M. K. Moe, *Phys. Rev. Lett.* 59 (1987), 2020. Copyright 1987 by the American Physical Society.

are illustrated in Figure 9.11. Here the measured electron spectrum is shown in comparison to the calculated energy spectrum for double β-decay. Although the statistics are less than ideal there is clear evidence of agreement between the calculated and measured spectra. The total decay rate as observed in this experiment is consistent with the estimate based on geological observations.

Another situation can occur when single β-decay is energetically favorable but is unlikely because the spins and parities of the parent and daughter make the transition highly forbidden. In such a case an allowed double β-decay might occur. Several such situations may be identified but the most notable is the possible double β-decay of ^{48}Ca to ^{48}Ti. The energy-level diagram is illustrated in Figure 9.12. The ground state β-decay of ^{48}Ca to ^{48}Sc has a Q value of 0.28 MeV; the transitions to the excited states are correspondingly less. On the basis of spins and parities it can be seen that the ground state transition is a sixth forbidden decay and the excited state transitions are fourth forbidden decays. A possible alternative to these highly forbidden, low energy, single β-decays might be a higher energy, allowed double β-decay:

$$^{48}\text{Ca} \rightarrow {}^{48}\text{Ti} + 2e^- + 2\bar{\nu}_e. \tag{9.47}$$

Here the transition is allowed, as it is $0^+ \rightarrow 0^+$, and has a Q of 4.27 MeV. Some recent experimental results of the double β-decay electron energy spectrum from ^{48}Ca are shown in Figure 9.13 along with the theoretically calculated

Figure 9.12 | Possible Double β-decay of ^{48}Ca

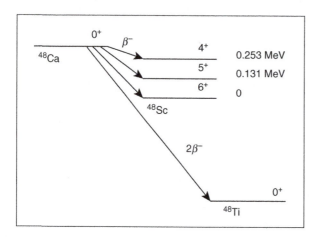

curve. In general there is reasonable agreement shown and the measured life-time, 6.2×10^{19} years, is consistent with theoretical predictions.

The number of double β-decays that have been detected experimentally is quite small because of the long lifetimes involved. However, experimental evidence as shown in Figures 9.11 and 9.13 taken in conjunction with geological

Figure 9.13 | Measured (data points) and Calculated (solid line) Electron Energy Spectrum from the Double β-decay of ^{48}Ca

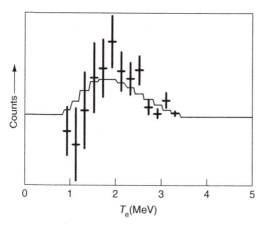

Data are from A. Balysh *et al.*, *Phys. Rev. Lett.* **77** (1996), 5186. Copyright 1996 by the American Physical Society.

data offers clear evidence for double β-decay. Additional ideas on double β-decay and some related processes will be presented in Chapter 17.

Problems

9.1. Describe all possible β-decay modes for ^{64}Cu.

9.2. Describe all possible decay schemes for tritium. Which of these is/are energetically favorable? Which is/are observed to occur?

9.3. (a) Using known masses determine the Q for the β-decay of ^{191}Os.
(b) Electrons from the β-decay of ^{191}Os are observed experimentally. What is their maximum energy?

9.4. Determine the degree to which the following ground state to ground state β-decays are forbidden: ^{10}Be, ^{21}F, ^{37}S, and ^{60}Co.

9.5. For odd A nuclei it is normally expected that only one β-stable nuclide will exist for each value of A. The ground states of ^{113}Cd and ^{113}In are both considered stable.
(a) Is it energetically favorable for one of these nuclides to decay to the other?
(b) If one of these decays is favorable why does it not occur?
(c) What is the relevance of the fact that the first excited state of ^{113}In has an energy of 0.393 MeV and a spin and parity of $1/2^-$?

9.6. Locate information about the β^--decay of ^{43}K from the *Table of Isotopes* or another suitable source. Describe as quantitatively as possible the reasons for the branching ratios to the various states of the daughter nuclide.

9.7. The endpoint energy for the electrons produced in the β^--decay of ^{8}Li is 13.1 MeV. Explain.

9.8. Show that ^{9}B may decay by proton emission, β^+-decay and electron capture but not by neutron emission or β^--decay.

9.9. Locate information about the excited states of ^{46}Ti. Discuss the possibility of β^--decay of the ground state of ^{46}Sc.

9.10. Derive expressions for the Q and barrier height for double β-decay in terms of atomic masses.

10 CHAPTER | Gamma Decay

10.1 ENERGETICS OF GAMMA DECAY

A nucleus in an excited state can decay to a lower energy (lower mass) state by gamma (γ) emission or internal conversion. The transition can be between an excited state and another lower energy excited state or between an excited state and the ground state. Such excited states can be readily formed in the daughter nucleus of an α- or β-decay process or by various nuclear reactions as discussed in detail in the next chapter. In the simplest case we can draw an analogy between nuclear transitions between single particle states and atomic transitions between electronic energy levels. In the former case electromagnetic radiation is produced in the form of a γ-ray, in the latter case an x-ray or optical photon is emitted. Internal conversion is a process by which γ-decay energy liberates an atomic electron rather than a photon and is analogous to an Auger process for atomic transitions. For electronic transitions it is easy to see how transitions involving charged electrons can emit radiation and the analogy can be drawn with single particle nuclear states involving a charged proton. However, single particle nuclear transitions involving a neutron also produce radiation and in a classical sense this can be related to the fact that the neutron, although uncharged, carries a magnetic moment (which in turn is related to the neutron's internal structure). Many nuclear transitions are not between single particle states but involve multiple nucleon excitations, rotational states, or vibrational states.

The energetics of γ-decay can be described in terms of the masses of the initial and final states, M_i and M_f, respectively, as

$$M_i c^2 = M_f c^2 + E_\gamma + E_R \tag{10.1}$$

where E_γ is the energy of the emitted γ-ray and E_R is the recoil energy of the nucleus. Conservation of linear momentum requires that

$$p_R = \frac{E_\gamma}{c} \qquad (10.2)$$

where p_R is the momentum of the recoiling nucleus and E_γ/c is the momentum carried by the γ-ray. From this equation the nuclear recoil energy is expressed as

$$E_R = \frac{E_\gamma^{\,2}}{2M_f c^2}. \qquad (10.3)$$

This derivation is nonrelativistic and the recoil energy is sufficiently small to justify this assumption. Using a typical γ-decay energy of 1 MeV and a nuclear mass of $A = 100$ gives a recoil energy of about 5 eV. This is much smaller than the γ-ray energy and, to within the accuracy of most direct γ-ray energy spectrum measurements, this can be ignored. Thus we can generally approximate

$$E_\gamma = (M_i - M_f)c^2. \qquad (10.4)$$

However, it is important to note that this recoil energy actually decreases the γ-ray energy by the amount in equation (10.3) and that this is much larger than the typical Heisenberg line width of an excited nuclear state as given by equation (7.27). As a result a consideration of the recoil energy can be extremely important for γ-resonance experiments such as Mössbauer effect spectroscopy.

10.2 CLASSICAL THEORY OF RADIATIVE PROCESSES

A simple model of a γ-decay process can be derived on the basis of classical electrodynamics. A static electric field is produced by a distribution of charges and as discussed in detail in Chapter 6, the charge distribution can be described in terms of a multipole expansion. A time varying distribution of charges produces a time varying electric field and this gives rise to the emission of radiation. When the time variation of the electric field is periodic (for example, sinusoidal) then a radiation field at the same frequency, ω, is produced. Like the static case, this radiation field can be described in terms of a multipole expansion. For example, consider the lowest order multipole term arising from an electric dipole. A simple model of an electric dipole consists of equal positive and negative charges separated by a distance a as shown in Figure 10.1 and defines an electric dipole moment

$$\vec{d} = q\vec{a} \qquad (10.5)$$

along the z-axis. A radiation field at a frequency ω can be produced by allowing the dipole to oscillate along the z-axis so that

$$\vec{d}(t) = q\vec{a}\sin\omega t. \qquad (10.6)$$

Figure 10.1 | The Electric Dipole

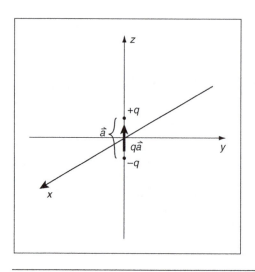

A similar situation can be described by the multipole expansion of the magnetic moments of a system. These moments can be modeled on the basis of currents and in the simplest case the magnetic dipole moment may be thought of as arising from a single loop of current. Figure 10.2 shows a magnetic dipole moment, $\vec{\mu}$, formed by a current, i, enclosing an area $\vec{A}$ where

$$\vec{\mu} = i\vec{A}. \tag{10.7}$$

A sinusoidally varying the current produces a time varying magnetic dipole moment of the form

$$\vec{\mu}(t) = i\vec{A}\sin\omega t \tag{10.8}$$

and a corresponding radiation field at frequency ω.

The basic properties of the radiation fields as described above can be determined by the application of classical electrodynamics. Of particular relevance to our future discussions is a calculation of the total radiated power. For the electric dipole this is found to be

$$P_e = \frac{1}{12\pi\varepsilon_0}\frac{\omega^4 d^2}{c^3}. \tag{10.9}$$

For the magnetic dipole this is

$$P_m = \frac{\mu_0}{12\pi}\frac{\omega^4 \mu^2}{c^3}. \tag{10.10}$$

Figure 10.2 | The Magnetic Dipole

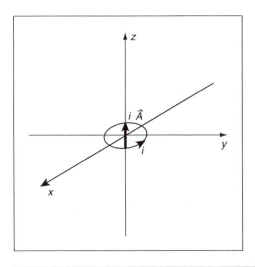

Here the total radiated power represents the energy emitted in the form of radiation per unit time.

For a generalized charge and/or current distribution, higher order multipole moments exist as discussed in Chapter 6. For example, an electric quadrupole moment can be constructed from four charges as illustrated in Figure 10.3. The order of the multipole moment, L, is defined as $L = 1$ for the dipole, $L = 2$ for the quadrupole, and so on. In general, the power radiated by an electric multipole moment of order L oscillating at a frequency ω can be calculated as

$$P_e(L) = \frac{2(L+1)c}{\varepsilon_0 L[(2L+1)!!]^2}\left(\frac{\omega}{c}\right)^{2L+2} Q_L^{\,2}, \qquad (10.11)$$

Figure 10.3 | The Electric Quadrupole

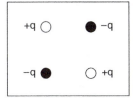

where Q_L is the generalized electric multipole moment of order L and the double factorial is defined for odd or even arguments such as $7!! = 1 \times 3 \times 5 \times 7$ or $8!! = 2 \times 4 \times 6 \times 8$, respectively.

Similarly for the magnetic case:

$$P_m(L) = \frac{2(L+1)\mu_0 c}{L[(2L+1)!!]^2} \left(\frac{\omega}{c}\right)^{2L+2} M_L^{\,2}, \tag{10.12}$$

where M_L is the generalized magnetic multipole moment of order L. The definition of Q_L and M_L for $L = 1$ differs from the definitions given in equations (10.5) and (10.7) only by some constant numerical factors.

The parity of the radiation field can be understood by investigating the effect of the transformation $r \rightarrow -r$ on the corresponding multipole moment. For example, for the electric dipole of Figure 10.1, the application of this transformation results in $d \rightarrow -d$ and the radiation field is defined as having odd parity (-1). On the other hand, application of the transformation $r \rightarrow -r$ to the magnetic dipole shown in Figure 10.2 causes no change in the resulting μ. This radiation field is defined as having even parity $(+1)$. This procedure can be applied to the electric quadrupole shown in Figure 10.3 leading to no change in the electric quadrupole moment and even parity. In general, the parity of an electric multipole radiation field of order L is given as

$$\pi_L = (-1)^L \tag{10.13}$$

and for a magnetic multipole field:

$$\pi_L = (-1)^{L+1}. \tag{10.14}$$

The radiation produced by an electric multipole is designated in terms of the value of L as EL. Magnetic mutipole radiation is referred to as ML. Table 10.1 gives the designations and parities of the first few orders of multipole radiation.

Table 10.1 | Properties of Some Multipole Radiation Types

Type	Symbol	L	Parity
electric dipole	E1	1	-1
magnetic dipole	M1	1	$+1$
electric quadrupole	E2	2	$+1$
magnetic quadrupole	M2	2	-1
electric octupole	E3	3	-1
magnetic octupole	M3	3	$+1$
electric hexadecapole	E4	4	$+1$
magnetic hexadecapole	M4	4	-1

In order to relate these ideas to nuclear γ-decay, it is necessary to consider them in a quantum mechanical context as discussed in the next section.

10.3 QUANTUM MECHANICAL DESCRIPTION OF GAMMA DECAY

Quantum mechanically the radiation given off by a varying electric or magnetic multipole is in the form of photons, each with energy $E_\gamma = \hbar\omega$. Thus the decay constants are related to the power radiated, as given in equations (10.11) and (10.12), by the simple expressions

$$\lambda_e(L) = \frac{P_e(L)}{\hbar\omega} \quad \text{and} \quad \lambda_m(L) = \frac{P_m(L)}{\hbar\omega} \tag{10.15}$$

for the electric and magnetic cases, respectively. Equations (10.11) and (10.12) can be used, more or less, for the quantum mechanical calculation of the decay constants except that the multipole moments must be replaced by appropriate multipole operators.

Looking first at the electric multipole case, we can write

$$\lambda_e(L) = \frac{2(L+1)}{4\pi\varepsilon_0 \hbar L[(2L+1)!!]^2} \left(\frac{E_\gamma}{\hbar c}\right)^{2L+1} \left|Q_{if}(L)\right|^2 \tag{10.16}$$

where

$$Q_{if}(L) = \int \psi_f^* Q(L) \psi_i \, d^3\vec{r}. \tag{10.17}$$

Following along the ideas presented in the last chapter on decay probabilities, we have written the decay probability as a function of the matrix element $Q_{if}(L)$, which is defined in terms of an electric multipole operator, $Q(L)$. The electric multipole operator is found to be proportional to $r^L Y_{LM}^*(\theta, \phi)$; see Section 5.3. As in the discussion of β-decay, ψ_i and ψ_f are the wave functions of the initial and final states, respectively. In general, the calculation of the matrix element in (10.17) is not straightforward. However, some rough estimates (referred to as the Weisskopf estimates) can be made on the basis of assumptions that are suitable for many cases. These assumptions are

1. The initial and final states are given by the single particle wave functions $\Psi_i = R_i(r) \, Y_{LM}(\theta, \phi)$ and $\Psi_f = R_f(r)$. The wave function indicates that the final state is assumed to be an s-state.
2. The radial part of the wave functions, $R_i(r)$ and $R_f(r)$, are constant over the nuclear volume and zero outside the nucleus.

Equation (10.17) can then be written as the integral over the nuclear radius R_0:

$$Q_{if}(L) = \frac{e\int_0^{R_0} r^L d^3\vec{r}}{\int_0^{R_0} d^3\vec{r}} \qquad (10.18)$$

where the integration in the denominator is needed for normalization purposes. Using $d^3\vec{r} = 4\pi r^2 dr$ this is immediately integrated to yield

$$Q_{if}(L) = e\left[\frac{3}{L+3}\right]R_0^{\ L} \qquad (10.19)$$

where R_0 can be related to A as in equation (3.23); $R_0 = R_1 A^{1/3}$. Equation (10.16) can now be written as

$$\lambda_e(L) = \frac{2e^2(L+1)}{4\pi\varepsilon_0\hbar L[(2L+1)!!]^2}\left[\frac{3}{L+3}\right]^2\left(\frac{E_\gamma}{\hbar c}\right)^{2L+1} R_1^{\ 2L} A^{2L/3}. \qquad (10.20)$$

This expression can be evaluated numerically for different L as a function of E_γ and A and is summarized in Table 10.2.

In the case of magnetic transitions a similar approach can be taken where the electric multipole operator in equations (10.16) and (10.17) is replaced with the corresponding magnetic multipole operator

$$M_{if}(L) = \int \psi_f^* M(L)\psi_i\, d^3\vec{r}. \qquad (10.21)$$

The magnetic multipole operator is found to be proportional to $r^L Y_{LM}^*(\theta, \phi)\nabla \cdot (\vec{r} \times \vec{j})$ where $\vec{j}$ is the current density inside the nucleus. It is also necessary to include a term to account for the intrinsic spin of the unpaired nucleon.

Table 10.2	Energy and A Dependence of Multipole Decay Probabilities. The γ-Ray Energy Is Expressed in MeV

L	$\lambda_e(L)$ (s^{-1})	$\lambda_m(L)$ (s^{-1})
1	$1.02 \times 10^{14}\, A^{2/3} E_\gamma^3$	$3.15 \times 10^{13}\, E_\gamma^3$
2	$7.28 \times 10^7\, A^{4/3} E_\gamma^5$	$2.24 \times 10^7\, A^{2/3} E_\gamma^5$
3	$33.9\, A^2 E_\gamma^7$	$10.4\, A^{4/3} E_\gamma^7$
4	$1.07 \times 10^{-5}\, A^{8/3} E_\gamma^9$	$3.27 \times 10^{-6}\, A^2 E_\gamma^9$

After some approximations, the transition rate for the magnetic case as obtained from the Weisskopf estimates is

$$\lambda_m(L) = \frac{20e^2\hbar(L+1)}{4\pi\varepsilon_0 c^2 m_p^2 L[(2L+1)!!]^2} \left[\frac{3}{L+3}\right]^2 \left(\frac{E_\gamma}{\hbar c}\right)^{2L+1} R_1^{2L-2} A^{(2L-2)/3}. \quad (10.22)$$

Numerically it is found from equations (10.20) and (10.22) that $\lambda_m(L) = 0.308 \cdot A^{-2/3}\lambda_e(L)$. These expressions provide the estimates of the decay constants as given in Table 10.2.

The predictions summarized in Table 10.2 should not be taken too seriously as the assumptions that have been made do not allow for the calculation of precise numerical values. However, these results are useful in understanding the relative decay rates for E and M radiation as a function of L and can be summarized as follows.

1. For a given transition energy there is a substantial decrease in decay constant with increasing L.
2. Electric transitions have decay constants that are typically about two orders of magnitude higher than for the corresponding magnetic transition constant.

The quantum mechanical selection rules, as discussed in the next section, must be considered in order to determine which of these transitions will actually occur.

10.4 SELECTION RULES

Conservation of angular momentum requires that the total angular momentum of the photon, given by L in the above discussion, be related to the total angular momentum of the initial and final nuclear states as

$$\left|\vec{J}_i - \vec{J}_f\right| \le L \le \left|\vec{J}_i + \vec{J}_f\right| \qquad \text{with } L = 1, 2, 3, \dots. \quad (10.23)$$

This expression constrains the possible values of L for the transition on the basis of the properties of the initial and final states (although it puts no restrictions on whether the transition can be electric or magnetic). It is, however, also necessary to consider the question of parity. The parity of the photon for different types of radiation is given in Table 10.1. Conservation of parity requires that a $\pi = -1$ photon may be emitted only when there is a change in the parity of the nucleus and a $\pi = +1$ photon may be emitted only when there is no change in the parity of the nucleus. This selection rule in conjunction with the conservation of angular momentum specifies what kinds of transitions can occur between certain nuclear states. It should be noted that these selection rules do not allow $0^+ \rightarrow 0^+$ transitions since equation (10.23) can only be satisfied for $L = 0$, which is not allowed because the photon has an intrinsic spin of 1. There are a small number of cases where there is a 0^+ ground state and a 0^+ first excited state.

Table 10.3 | Examples of Allowed γ-Transitions

J_i^π	J_f^π	Nuclear Parity Change	L	Allowed Transitions
0^+	0^+	no	—	none
$1/2^+$	$1/2^-$	yes	1	E1
1^+	0^+	no	1	M1
2^+	0^+	no	2	E2
$3/2^-$	$1/2^+$	yes	1, 2	E1, M2
$5/2^+$	$1/2^-$	yes	2, 3	M2, E3
2^+	1^+	no	1, 2, 3	M1, E2, M3
$3/2^-$	$5/2^+$	yes	1, 2, 3, 4	E1, M2, E3, M4
$5/2^+$	$3/2^-$	no	1, 2, 3, 4	M1, E2, M3, E4

^{40}Ca is one example; recall the discussion of rotational levels in even–even nuclei in Section 6.7. In such cases the transition from the first excited state to the ground state must proceed by internal conversion (see below).

Often the above selection rules allow for more than one kind of decay. Some examples are given in Table 10.3. The relative importance of the various allowed decay modes can be estimated on the basis of equations (10.20) and (10.22). These equations allow for a determination of partial decay constants (or partial lifetimes) and a calculation of the branching ratios for the various decay modes. From the figure it is clear that the lowest order multipolar radiation will dominate. In many cases, for example a low energy E1 + M2 transition (for example, $3/2^-$ to $1/2^+$) the lowest order mode dominates and all other modes are negligible. In some cases, for example a fairly high energy M2 + E3 transition (for example, $5/2^+$ to $1/2^-$), the contribution from the second mode can be significant.

10.5 INTERNAL CONVERSION

For the decay from an excited state to a lower energy state internal conversion competes with γ-emission. For the $0^+ \rightarrow 0^+$ transition, that is an E0 transition, γ-decay is not allowed and internal conversion is the only possibility for de-excitation. Internal conversion refers to the process by which the de-excitation of nuclear energy levels gives up its energy to an atomic electron. If the energy involved, E_γ, is greater than the binding energy of the electron, b_n (which it most commonly is), the electron is liberated with a kinetic energy

$$T_e = E_\gamma - b_n. \tag{10.24}$$

Figure 10.4 | Electronic Binding Energies for Different Electron Shells

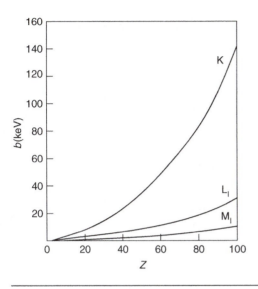

This expression conserves energy and a consideration of the total angular momentum of the electron relative to the initial and final nuclear state J is necessary to insure that total angular momentum is conserved. It is electrons in the inner shells (specifically s-orbital electrons) that are involved in this process because it is these electrons that have wave functions that have largest values at the nucleus. Typically only internal conversion electrons from K, L, and M shells are observed. The binding energy of these electrons is given as a function of Z in Figure 10.4. Thus the internal conversion electron spectrum consists of several peaks corresponding to different electron shells. An example is shown in Figure 10.5. The relative intensity of the peaks is related to the probability of internal conversion with electrons from the different shells and the figure suggests that shells beyond M will contribute very little to the spectrum. Although it is straightforward to observe the energy spectrum of the electrons emitted by internal conversion, the interpretation of the resulting spectrum is not usually as straightforward as suggested in Figure 10.5. If more than one excited state is involved then there will be transitions with different values of E_γ in equation (10.24), resulting in multiple sets of peaks. As well, if the excited state has been populated by a β^--decay then the internal conversion electron spectrum will be superimposed on the β-decay spectrum.

The total decay constant for γ-decay and internal conversion is given as the sum of the individual components

$$\lambda = \lambda_\gamma + \lambda_e \tag{10.25}$$

Figure 10.5 | Internal Conversion Electron Spectrum for the 1.416 MeV E0 Transition in ^{214}Po

Increasing magnetic field, from right to left, corresponds to increasing electron energy.

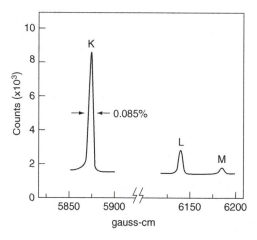

From D. E. Alburger and A. Hedgran, *Arkiv Fyzik 7* (1953), 424. Used with permission.

where the γ and e subscripts refer to γ-decay and internal conversion, respectively. The internal conversion decay constant can also be expanded in terms of the contributions from the K, L, M, ... shells as

$$\lambda_e = \lambda_K + \lambda_L + \lambda_M + \cdots. \quad (10.26)$$

The internal conversion coefficient is defined as the ratio of the total number of decays for a particular transition that proceed by internal conversion to those that proceed by γ-emission:

$$\alpha = \frac{\lambda_e}{\lambda_\gamma}. \quad (10.27)$$

The K-shell internal conversion coefficient can be described as

$$\alpha_K = \frac{\lambda_K}{\lambda_\gamma} \quad (10.28)$$

and similarly for L, M,... etc. giving the total internal conversion coefficient

$$\alpha_e = \alpha_K + \alpha_L + \alpha_M + \cdots. \quad (10.29)$$

The calculation of the internal conversion decay constants is not simple but follows along the lines of the discussion above concerning the determination of appropriate matrix elements for γ-emission. A nonrelativistic approach simplifies matters and although it cannot be expected to give accurate numerical results at higher energies, it provides some insight into the relevance of the important parameters. Internal conversion coefficients for electric and magnetic transitions of order L can be found in this way to be

$$\alpha(EL) = \frac{Z^3}{n^3}\left(\frac{L}{L+1}\right)\left(\frac{e^2}{4\pi\varepsilon_0\hbar c}\right)^4\left(\frac{2m_ec^2}{E_\gamma}\right)^{L+5/2}\tag{10.30}$$

and

$$\alpha(ML) = \frac{Z^3}{n^3}\left(\frac{e^2}{4\pi\varepsilon_0\hbar c}\right)^4\left(\frac{2m_ec^2}{E_\gamma}\right)^{L+3/2}\tag{10.31}$$

where n is the principal quantum number corresponding to the electron energy level; $n = 1, 2, 3,...$ for K, L, M,... etc. A general conclusion based on these expressions indicates that the internal conversion coefficient is greatest for heavy elements (large Z), small transition energy (E_γ), and inner shell electrons (small n). Numerous graphs of internal conversion coefficients for electric and magnetic transitions of different multipolarities and for different values of A can be found in the *Table of Isotopes*.

Problems

10.1. Discuss the decay of the ground state and the first excited state of ^{121}Sn.

10.2. The spins and parities of the ground state and first four excited states of ^{83}Kr are $9/2^+$, $7/2^+$, $1/2^-$, $5/2^-$, and $3/2^-$, respectively. For all γ-transitions ending at the ground state determine the allowed multipolarities.

10.3. The validity of the Weisskopf estimate for the M2 transition may be demonstrated by plotting the energy dependence of $\tau A^{2/3}$ for decays of different energies. From the *Table of Isotopes* or another suitable reference locate four M2 transitions and show that such a plot is consistent with the Weisskopf estimate. Note that the following points may be of relevance:
(i) A logarithmic plot may best illustrate the features of the relationship.
(ii) Tables may give halflives rather than lifetimes.
(iii) In the case of multimodal decays it is important to use the correct lifetime for the partial decay.
(iv) It may be necessary to correct for the effects of internal conversion.

10.4. The first excited state of ^{58}Co has $J^\pi = 5^+$, $\tau = 12.0$ h and an energy of 0.025 MeV. For ^{58}Fe these values are 2^+, 10^{-11} s, and 0.81 MeV, respectively. Does the Weisskopf estimate provide a good description of the decays of these two excited states?

10.5. Describe the type and multipolarity of possible γ-decays between the following spin and parity states: (a) $9/2^- \rightarrow 7/2^+$, (b) $1/2^- \rightarrow 7/2^-$, (c) $4^+ \rightarrow 2^+$, and (d) $11/2^- \rightarrow 3/2^+$.

10.6. (a) Calculate the recoil energy for the transitions from the first excited state to the ground state for the following nuclides (energy given in parentheses): (i) ^{15}O (5.183 MeV), (ii) ^{19}O (0.0960 MeV), (iii) ^{57}Fe (0.0144 MeV), (iv) ^{70}Ge (1.0396 MeV), (v) ^{227}Th (0.0093 MeV), and (vi) ^{228}Th (0.0578 MeV). (b) Comment on the general trends that are expected for the importance of the recoil energy in terms of nuclear properties.

10.7. Calculate the Weisskopf value for the transition from the first excited state to the ground state for tungsten isotopes with $A = 180$, 182, 184, and 186. Compare these with experimentally measured values. What common features do these transitions have that could explain these results?

10.8. A nucleus has a $^{1}/_{2}^{-}$ ground state. The excited states are, more or less, evenly spaced in energy and have spins and parities of $^{5}/_{2}^{-}$, $^{3}/_{2}^{-}$, $^{7}/_{2}^{+}$, and $^{5}/_{2}^{+}$, respectively, with increasing energy. Draw an energy level diagram indicating all expected γ-transitions to the ground state, their multipolarities, and rough estimates of their expected decay rates.

Nuclear Reactions

11.1 GENERAL CLASSIFICATION OF REACTIONS AND CONSERVATION LAWS

A nuclear reaction is a process that results from the interaction between a nucleus and a particle that is incident upon it. In general, a nuclear reaction can be represented by a particle, a, incident on a nucleus, A, which produces another particle, b, and a resulting nucleus, B. This can be written as

$$a + A \rightarrow b + B. \tag{11.1}$$

A convenient shorthand notation that is in common use for the process in equation (11.1) is

$$A(a,b)B. \tag{11.2}$$

This convention will be used where appropriate throughout this chapter.

Nuclear processes can be roughly divided into two categories: *scattering* in which the incident particle and emitted particle are the same and *reactions* in which the incident and emitted particles are different. Scattering processes can be either elastic or inelastic. Elastic scattering resulting from coulombic interactions is referred to as coulombic or Rutherford scattering and has already been discussed in Chapter 3 in the context of experiments to determine the size of the nucleus. Such processes conserve kinetic energy. Inelastic scattering refers to the situation where kinetic energy is not conserved. In this case the scattering particle will lose kinetic energy and the nucleus can be left in an excited state.

141

Nuclear reactions may be *direct reactions, compound nucleus reactions,* or *resonance reactions.* A *direct reaction* is one in which the incident particle interacts only with a limited number of valence nucleons in the target nucleus. This situation is most likely when the de Broglie wavelength of the incident particle is comparable to the size of an individual nucleon (about 1 fm) rather than the size of the nucleus (about 10 fm). This happens when the energy of the incident particle is relatively high. A *compound nucleus reaction* is one in which the incident particle becomes bound to the nucleus forming a compound nucleus before the reaction continues. Thus the interactions between all the nucleons is important in determining the state of the compound nucleus. It is only the properties of the compound nucleus and not how it was formed that is of relevance in determining the way in which it decays. *Resonance reactions* are, in some ways, intermediate between these two extremes as the incident particle can become quasibound to the nucleus before the reaction proceeds. The distinction between these three situations is not always clear and in many cases a process can involve more than a single mode.

Several conservation laws must be considered when looking at the details of a reaction. Mass/energy and momentum must be conserved although kinetic energy, in many cases, is not. At sufficiently high energy new types of particles may be created (usually mesons). This can occur at energies above about 280 MeV, corresponding to the production of pions. These reactions are discussed further in Section 16.2. At lower energies the number and identity of particles is normally conserved. Specifically the number of neutrons and the number of protons will not change except in cases where the weak interaction is important. Charge must be conserved but conservation of the number of protons will insure this in any case. Here we will consider relatively low energy reactions that involve the strong or electromagnetic interaction. High-energy reactions in which other kinds of particles can be created will be considered in Part IV of this book. Low energy processes generally involve incident particles that are neutrons, protons, or bound systems consisting of a small number of nucleons. The most commonly encountered incident particles are summarized in Table 11.1.

Table 11.1 | Some Examples of Particles Involved in Nuclear Reactions

Symbol	Name	Nucleus of	Identity
n	neutron	—	neutron
p	proton	^{1}H	proton
d	deuteron	^{2}H	n + p
t	triton	^{3}H	2n + p
^{3}He	—	^{3}He	n + 2p
^{4}He	α-particle	^{4}He	2n + 2p

Because elastic scattering was considered previously, here, inelastic scattering and some different low energy reactions are discussed in detail.

11.2 INELASTIC SCATTERING

Consider the simple scattering event A(a, a)A* as illustrated in Figure 11.1 (the superscript "*" indicates that the nucleus is in an excited state). Here a particle scatters from a nucleus inelastically, losing energy and leaving the nucleus in an excited state. We will deal with this problem nonrelativistically as this is a good approximation for nucleons at the energies that we are considering. However, relativistic corrections can be included if necessary. The total energy in the laboratory frame before the collision, E, is given as the kinetic energy of the incident particle, E_i, and the rest mass energy of the system,

$$E = E_i + m_a c^2 + m_A c^2. \tag{11.3}$$

After the collision the total energy is

$$E = E_f + E_{A*} + m_a c^2 + m_{A*} c^2 \tag{11.4}$$

where E_f is the kinetic energy of the emitted particle and E_{A*} is the kinetic energy of the nucleus. The change in energy of the nucleus is represented by the change in its rest mass (that is, the energy of the excited state) and is found by equating (11.3) and (11.4) as

$$\Delta E = (m_{A*} - m_A)c^2 = E_i - E_f - E_{A*}. \tag{11.5}$$

This equation can be written in terms of the corresponding momenta as

$$\Delta E = \frac{p_i^2}{2m_a} - \frac{p_f^2}{2m_a} - \frac{p_{A*}^2}{2m_{A*}}. \tag{11.6}$$

Figure 11.1 | Geometry of an Inelastic Scattering Event

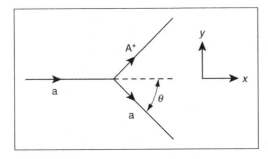

From the figure we can see that conservation of the x-component of momentum gives

$$p_{A^*x} = p_i - p_f \cos\theta \tag{11.7}$$

and conservation of the y-component of momentum gives

$$p_{A^*y} = p_f \sin\theta. \tag{11.8}$$

Substituting these expressions into equation (11.6) for p_{A^*} gives

$$\Delta E = \frac{p_i^2}{2m_a} - \frac{p_f^2}{2m_a} - \frac{1}{2m_{A^*}}\left[p_i^2 + p_f^2 - 2p_ip_f\cos\theta\right]. \tag{11.9}$$

In terms of initial and final energies of the scattered particle this becomes

$$\Delta E = E_i\left(1 - \frac{m_a}{m_{A^*}}\right) - E_f\left(1 + \frac{m_a}{m_{A^*}}\right) + 2\frac{m_a}{m_{A^*}}\sqrt{E_iE_f}\cos\theta. \tag{11.10}$$

Elastic scattering is seen to be a limiting case of this expression with $\Delta E = 0$ where, for a given scattering angle, θ, the difference between E_i and E_f accounts for the kinetic energy (recoil) of the nucleus. From equation (11.5) it is seen that in the case of inelastic scattering the actual mass of the nucleus after collision can only be calculated from equation (11.10) by an iterative process. However, to the accuracy of experimental measurements, the value of m_{A^*} on the right-hand side of equation (11.10) can be replaced by m_A, the known ground state mass. Since the atomic electrons are associated with the nucleus and there is no change in the identity of the nucleus, it is appropriate to use measured atomic masses in the analysis of the above equations to describe the dynamics of the interaction.

Experimentally the excited state energies of a nucleus can be investigated by allowing a beam of monoenergetic particles to be incident on the nucleus and measuring the energy spectrum of the particles scattered at a specific angle. As an example let's consider the inelastic scattering of 10.02 MeV protons from a sample containing ^{10}B nuclei, ^{10}B (p, p)^{10}B*. For a fixed scattering angle, say 90°, protons that have a final energy E_f will give up an energy ΔE to the ^{10}B nucleus. If ΔE is equal to the energy difference between the ^{10}B ground state and one of the ^{10}B excited states then the energy given up by the proton can cause an excitation of the ^{10}B nucleus. Since the nucleus has a number of well-defined excited states we expect that the protons scattered by a specific angle will have a series of well-defined energies, or resonances, related to the excited state energies by equation (11.10). The proton energy spectrum from such an experiment is shown in Figure 11.2. The left-most resonance corresponds to elastic scattering where the ^{10}B nucleus is left in the ground state. With decreasing energy each peak represents a larger ΔE and a higher energy excited state for the ^{10}B nucleus. An analysis on the basis of equation (11.10) gives the excitation energies in Table 11.2 and the energy levels for ^{10}B as shown in Figure 11.3.

Figure 11.2 | Number of 10.02 MeV Protons Scattered at 90° from a Sample Containing ^{10}B as a Function of Final Proton Energy

Decreasing proton energy, from left to right in the figure, corresponds to increasing ^{10}B excited state energy as shown in Table 11.2. The ^{10}B peaks along with their corresponding excited state energy are indicated.

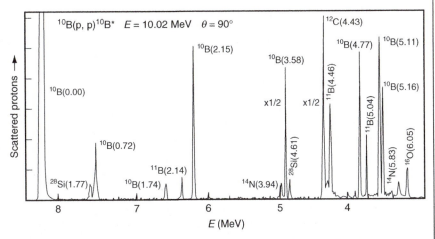

Reprinted from B. H. Armitage and R. E. Meads, "Energy levels of B^{10} (I) the reactions B^{10}(p, p')B^{10}, B^{10}(d, d')B^{10}, and C^{12}(d, α)B^{10}," *Nuclear Physics 33* (1962), 494–501. Copyright 1962, used with permission from Elsevier Science.

Table 11.2 | Relationship of Final Proton Energy to Excited State Energy for 10.02 MeV Protons Scattered from ^{10}B at an Angle of 90°

E_f (MeV)	ΔE (MeV)
8.19	0
7.53	0.72
6.61	1.74
6.23	2.15
4.93	3.59
3.85	4.77
3.54	5.11
3.50	5.18

Data are obtained from Figure 11.2.

Figure 11.3 | Energy Levels for ^{10}B

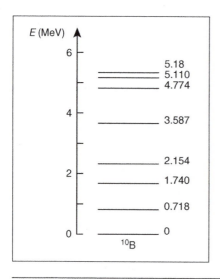

11.3 NUCLEAR REACTIONS

Nuclear reactions can also occur where the identity of the incident and emitted particles are different. Several different situations are possible and these may be described as follows:

1. An incident nucleon is absorbed by the nucleus leaving the nucleus in an excited state and the emitted particle is a γ-ray resulting from the de-excitation of the nucleus. A well known example is the (n, γ) reaction.
2. An incident nucleon is absorbed by the nucleus and a different nucleon is emitted. An example is the (n, p) reaction.
3. The incident particle loses one (or more) of its nucleons to the nucleus; for example, the (d, p) reaction. This is referred to as a *stripping reaction* and is discussed further in Section 11.4.
4. The incident particle gains one (or more) nucleon from the nucleus; for example, the (d, α) reaction. This is referred to as a *pick-up reaction*.
5. The reaction causes the component nucleons of the incident particle to become unbound. An example is the (d, np) reaction, where the terminology np refers to a neutron proton pair that is not bound.

As an example of some of the above reactions let's consider a deuteron incident on an ^{16}O nucleus. The possible reactions involving the particles given in Table 11.1 are as follows:

$$d + {}^{16}O \rightarrow {}^{16}O + d$$
$$d + {}^{16}O \rightarrow {}^{18}F$$

Figure 11.4 | Schematic Illustration of Q for Various $^{16}O + d$ Reactions

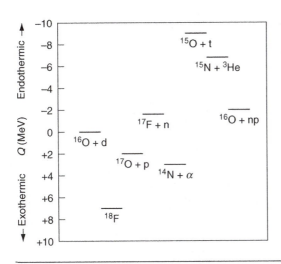

$$d + {}^{16}O \rightarrow {}^{17}O + p$$
$$d + {}^{16}O \rightarrow {}^{17}F + n$$
$$d + {}^{16}O \rightarrow {}^{14}N + \alpha \qquad (11.11)$$
$$d + {}^{16}O \rightarrow {}^{15}O + t$$
$$d + {}^{16}O \rightarrow {}^{15}N + {}^{3}He$$
$$d + {}^{16}O \rightarrow {}^{16}O + np.$$

The energetics of these processes can be readily established on the basis of measured atomic masses, although it is important to properly account for possible changes in the number of electrons associated with the nucleus before and after the reaction. A diagram indicating the Q for each of the reactions given in equation (11.11) is shown in Figure 11.4. The vertical energy scale is referenced to $^{16}O + d$. Reactions with positive Q have reaction products with a smaller mass than the initial components and reactions with negative Q have reaction products with a larger mass than the initial components. The former are referred to as exothermic and the latter as endothermic. Unlike decay processes, endothermic reactions can occur because additional kinetic energy can be supplied by the incident particle. It is the kinetic energy in the center of mass frame that is of importance in this problem. Since experimentally the incident particle energy is measured in the laboratory it is necessary to transform to the center of mass frame as:

$$E_{cm} = \frac{E_{lab}}{1 + \dfrac{m_a}{m_A}}. \qquad (11.12)$$

Thus the center of mass energy is always less than the measured laboratory energy.

Figure 11.4 shows that the (d, γ), (d, p), and (d, α) reactions are exothermic and the (d, n), (d, t), (d, ^{3}He), and (d, np) reactions are endothermic. These latter processes require a minimum center of mass kinetic energy of 1.62, 9.41, 6.63, and 2.22 MeV, respectively, in order to proceed. These centers of mass energies correspond to laboratory energies of 1.82, 10.59, 7.46, and 2.50 MeV, respectively.

Energy above the amount required to cause a reaction to proceed may yield additional kinetic energy of the reaction byproducts and/or may leave the resulting nucleus in an excited state. That is, any excited state at or below an energy

$$E = E_{cm} + Q \tag{11.13}$$

is accessible. This requires the existence of an excited state in this energy range in the resulting nucleus. In general we know that the density of excited states increases with increasing nuclear mass. It is also known that the density of excited states in a given nucleus generally increases with increasing energy. This is readily understood on the basis of the shell model. The lowest lying states are often single nucleon states. Higher energy states generally involve multiple nucleon excitations. As the number of nucleons participating increases so does the number of combinations of state occupancies that will give similar but slightly different excited state energies. Thus for heavy nuclei, and for cases where the available energy in equation (11.13) is fairly large (or both), the spacing (in energy) between the resonances is small. The practical implication of this will be seen in Chapter 12.

The angular dependence of the energy spectrum of emitted particles in a nuclear reaction may be calculated (nonrelativistically) in a manner similar to the derivation in Section 11.2. This calculation gives

$$\Delta E = E_i\left(1 - \frac{m_a}{m_B}\right) - E_f\left(1 + \frac{m_b}{m_B}\right) + 2\frac{\sqrt{m_a m_b E_i E_f}}{m_B}\cos\theta + Q \tag{11.14}$$

where the nucleus B may be left in an excited state. A particularly interesting and useful example of the application of this expression deals with the deuteron stripping reactions (d, n) and (d, p). In these processes the incident deuteron loses one of its nucleons and this is absorbed by the nucleus. This is an effective way of looking at excited states of the resulting nucleus and is discussed further in the next section.

11.4 DEUTERON STRIPPING REACTIONS

In the example of the (d, p) process in equation (11.11), deuteron stripping may be used to examine the excited states of ^{17}O. Since this reaction is exothermic, even low energy deuterons may excite any ^{17}O levels below about 1.9 MeV (although coulombic effects, as discussed in Section 11.6, are of some significance). Additional deuteron kinetic energy as measured in the center of mass frame may excite higher energy ^{17}O levels. As an example, the peaks in the energy spectrum of protons emitted by the ^{16}O(d, p)^{17}O reaction at an angle of 25° with respect to a beam of 10 MeV (laboratory frame) deuterons are summarized in

Table 11.3 | Final Proton Energy and Excited State Energy for the $^{16}O(d, p)^{17}O^*$ Reaction for a Scattering Angle of $25°$

E_f (MeV)	ΔE (MeV)
11.69	0.00
10.81	0.871
8.58	3.06
7.77	3.85
7.05	4.56
6.50	5.08
6.19	5.38
5.85	5.71
5.66	5.89
5.60	5.94

Table 11.3. Excited state energies as calculated using equation (11.14) are summarized in the table and are illustrated in Figure 11.5.

It is important to note in the figure the location of the neutron separation energy for ^{17}O at 4.15 MeV. Excited ^{17}O states above 4.15 MeV are sometimes thought of as n-^{16}O resonances as the neutron is only quasibound to the ^{16}O nucleus.

Figure 11.5 | Energy Level Diagram for ^{17}O

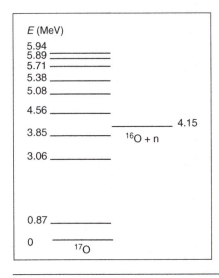

Since ^{17}O is β-stable, excited states below 4.15 MeV may decay only by γ-decay (or internal conversion, as appropriate). However it is energetically favorable for excited states above 4.15 MeV to decay either by γ-decay or by neutron emission. The interaction between the neutron and the ^{16}O nucleus can be understood better by looking at the reactions involving neutrons incident upon nuclei as discussed in the next section.

11.5 NEUTRON REACTIONS

A neutron that is incident on an ^{16}O nucleus (for example) may be absorbed forming ^{17}O. The energy available in this process is given by equation (11.13) and in this example is the center of mass energy of the neutron plus 4.15 MeV. When the available energy is equal to the energy of an ^{17}O excited state then a resonant condition exists and the neutron absorption cross section is greatly enhanced. The measured neutron absorption cross section for ^{16}O as a function of center of mass energy above 0.6 MeV is illustrated in Figure 11.6. The resonant peaks correspond to the formation of ^{17}O excited states. Only those states above the ^{17}O neutron separation energy of 4.15 MeV are accessible as this is the Q available from a reaction with a very low energy neutron. The resonant peaks in Figure 11.6 may be related to the excited state energy levels of ^{17}O as shown in Figure 11.5. For example, the lowest energy peak in Figure 11.6 at 0.93 MeV corresponds to the ^{17}O excited state at $0.93 + 4.15 = 5.08$ MeV. Once ^{17}O has been formed in an excited state above 4.15 MeV it may decay by neutron emission or by γ-decay. If the decay is by γ-decay to an energy level below 4.15 MeV, then the ^{17}O nucleus can no longer decay by neutron emission and the neutron is said

Figure 11.6 | Total Neutron Cross Section for ^{16}O as a Function of Center of Mass Energy

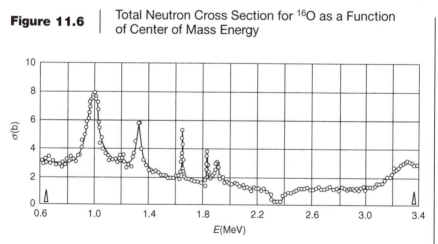

From D. J. Hughes and R. B. Schwartz, *Neutron Cross Sections*, second edition. Upton, NY: Brookhaven National Laboratory, 1958. Courtesy of Brookhaven National Laboratory.

Figure 11.7 | (n, γ) and (n, n) Reaction Channels for a Neutron Incident on ^AX

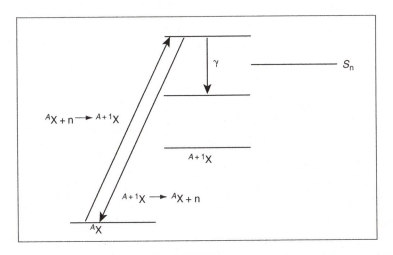

to have been captured. This process is the so-called (n, γ) reaction and has important applications for the study of excited state structure, the formation of radioactive nuclides, and chemical analysis by neutron activation analysis.

It is interesting to note the values of the cross sections as shown in Figure 11.6. The radius of an ^{16}O nucleus is about 3 fm, giving a cross sectional area of about 0.3 barns (30 fm^2). The figure shows cross sections that are nearly two orders of magnitude larger. In fact (n, γ) cross sections for some nuclei may be in excess of 10^5 barns. Thus, although the cross section is expressed in units of area it should not be viewed as the physical cross sectional area. Rather, the cross section results from the quantum mechanics of the interaction between the neutron and the nucleus and is a measure of the reaction probability.

In addition to providing information about the energies of the excited states, the energy dependence of the neutron cross section can provide information about the stability of these states. In Chapter 7 it was shown that the energy width of an unstable state is related to its lifetime by the Heisenberg uncertainty principle. In the example illustrated in Figure 11.7, a neutron is absorbed by the nucleus ^AX forming an excited state of ^{A+1}X at an energy above the neutron separation threshold of ^{A+1}X. This state may decay by two modes (which are referred to as channels); γ-decay to a lower energy level of ^{A+1}X or neutron emission back to ^AX. This latter is referred to as the incident channel for the reaction. Following along the lines of equations (7.12) and (7.26), the total width, Γ, of the ^{A+1}X excited state is the sum of the partial widths for decay by γ-emission, Γ_γ, and decay by the incident channel, Γ_i:

$$\Gamma = \Gamma_\gamma + \Gamma_i. \tag{11.15}$$

From the line shape given in equation (7.25), we expect an energy dependent cross section for neutrons incident upon ^AX with an energy in the vicinity of the ^{A+1}X excited state energy E_0;

$$\sigma_i(E) = \frac{C}{(E - E_0)^2 + \frac{\Gamma^2}{4}} \qquad (11.16)$$

where the constant C is related to the density of states in the system. This is found in the following way. The reaction is assumed to take place within the volume of the nucleus, V. Following along the lines of the discussion in Section 9.2, the density of states per unit momentum for fermions within this volume is given by equation (9.23) as

$$\frac{dn}{dp} = \frac{4\pi p^2}{(2\pi\hbar)^3} V. \qquad (11.17)$$

If the neutron, moving with velocity v within the volume V, has a cross sectional area σ_i it sweeps out a volume $\sigma_i v$ per unit time. The decay rate per unit momentum is then given as the fraction of the volume V swept out per unit time multiplied by the density of states:

$$d\lambda = \frac{\sigma_i v}{V} dn = \sigma_i v \frac{4\pi p^2}{(2\pi\hbar)^3} dp. \qquad (11.18)$$

This gives the total decay rate by integrating over momentum

$$\lambda = \frac{4\pi}{(2\pi\hbar)^3} \int_{-\infty}^{+\infty} v\sigma_i p^2 dp. \qquad (11.19)$$

Substituting (11.16) for the cross section and changing the integration over momentum to an integration over energy gives

$$\lambda = \frac{8\pi m EC}{\Gamma} \qquad (11.20)$$

where m is the reduced mass of the system:

$$m = \frac{m_n m_X}{m_n + m_X}. \qquad (11.21)$$

In a system with a large number of neutrons and nuclei the rate of formation of the excited state would be balanced by the rate of decay back through the incident channel. This is given as

$$\lambda = \frac{\Gamma_i}{\hbar}. \qquad (11.22)$$

Equating (11.20) and (11.22) yields an expression for C and the cross section from equation (11.16) becomes

$$\sigma_i(E) = \frac{1}{8\pi m E \hbar^2} \frac{\Gamma_i \Gamma}{(E - E_0)^2 + \dfrac{\Gamma^2}{4}}. \qquad (11.23)$$

A more detailed analysis of this problem shows that Γ must be replaced by $g\Gamma$ where the statistical factor g is given in terms of the spins of the incident neutron, s_n, and target nucleus, s_X, and the total angular momentum of the excited state, J;

$$g = \frac{2J + 1}{(2s_n + 1)(2s_X + 1)}. \qquad (11.24)$$

The above expression for the cross section is known as the Breit-Wigner formula and allows for the analysis of the resonance line shapes in data such as that shown in Figure 11.6.

The (n, γ) process is important for low energy neutrons as it is then typically the energy levels near or below the neutron separation energy that are populated directly leaving γ-decay as the likely decay mode. The relative scale of neutron energies depends on the nature of the target nucleus. In the case of neutrons incident on ^{16}O the first available energy level in ^{17}O above the neutron separation energy is at 4.56 MeV. Thus a center of mass neutron energy of 0.41 MeV is required to populate this level. On the average the neutron energy necessary to populate the first available state above the neutron separation energy is inversely proportional to the density of states at the energy Q. For reasons described previously, we expect, on the average, that the density of states will be greater when the value of Q places the resulting nucleus well above the ground state in energy and when the value of A is large. An example of this is illustrated in Figure 11.8. It is seen here that the location of the first resonance peak and the spacing between the peaks for ^{238}U is of the order of tens of eV, compared with 100's of keV for ^{16}O as seen in Figure 11.6.

Neutrons for experiments can be produced by a variety of neutron-emitting sources. Unlike charged particles (for example, electrons, protons, deuterons) neutrons cannot be accelerated by means of an electric field. However, higher energy neutrons as produced by a radioactive source, can be slowed to produce a source of lower energy neutrons. This is done by allowing the neutrons to lose kinetic energy through a variety of scattering mechanisms in an appropriate material, called a moderator. In general neutrons are classified according to their kinetic energy as summarized in Table 11.4. The ability to reduce the kinetic energy of a neutron from high energy (a few MeV) to a few eV or less is of crucial importance to the operation of a fission reactor and will be discussed in detail in the next chapter.

Neutron cross sections at high energies can be quite complex. At higher energies the increasing density of states decreases the average distance between resonance peaks. As the distance between the resonances becomes comparable to their halfwidth, the resonances overlap and at some point the concept of distinct energy levels becomes ambiguous. This will be discussed further in the next chapter with specific reference to ^{235}U and ^{238}U.

Figure 11.8 | Total Neutron Cross Section for ^{238}U as a Function of Center of Mass Energy

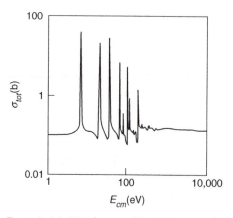

From A. M. Weinberg and E. P. Wigner, *The Physical Theory of Neutron Chain Reactions*. Chicago: University of Chicago Press, 1958. Copyright 1958 by University of Chicago Press. Used with permission.

An important point to consider when dealing with low energy neutron reactions is that of thermal effects. The thermal motion of the atoms in the target will influence the center of mass energy of the incident neutrons. If the motion of a neutron is described by a velocity $\vec{v}_n$ and the motion of an atom is described by a velocity vector $\vec{v}_A$, then the center of mass energy is given by

$$E = \frac{1}{2}m(\vec{v}_n - \vec{v}_A)^2 \tag{11.25}$$

where m is the reduced mass. This may be written in terms of the center of mass energy of the neutron, E_{cm}, and the thermal energy of the atom, $E_A = k_B T$ as

$$E = E_{cm} + \frac{m}{m_A}E_A - 2\left[\frac{m}{m_A}E_{cm}E_A\right]^{1/2}\cos\theta. \tag{11.26}$$

Table 11.4 | Categories of Neutrons Based on Their Kinetic Energy

Class	Typical E
thermal	0.02 eV
epithermal	1 eV
slow	1 keV
fast	>100 keV

Figure 11.9 | Effects of Doppler Broadening on the (n, γ) Resonance at 6.67 eV in ^{238}U

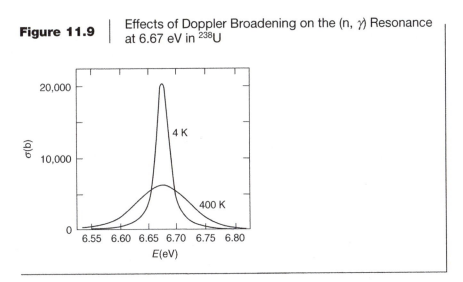

In general m/m_A will be small and except in the case of very low energy neutrons, E_A will be small compared to E_{cm}. Thus the second term on the right-hand side of equation (11.26) can normally be ignored. The $\cos\theta$ can take on values from -1 to $+1$ so that the actual energy may be reduced or increased relative to the value of E_{cm}. This results in a broadening of the resonance peak, the so-called Doppler broadening, where the width of the distribution of $\cos\theta$ values is approximately unity. Thus we can write the increased width due to thermal effects as

$$\Delta E = 2\sqrt{\frac{m}{m_A}E_{cm}k_BT}. \tag{11.27}$$

An example of the thermal broadening of a low energy neutron resonance in ^{238}U is shown in Figure 11.9.

11.6 COULOMBIC EFFECTS

In contrast to neutron reactions, reactions that involve charged particles are affected by coulombic interactions. This is true both in the case when the incident particle is charged and when the emitted particle is charged. The interaction potential between the charged nucleus and the charged particle forms a Coulomb barrier as for the case of α-decay described in Chapter 8. Unless the energy of the particle is greater than the barrier height, the calculation of reaction cross sections must be considered as a quantum mechanical tunneling problem. At low energies, charged particles that are incident on a nucleus predominantly undergo elastic scattering because of their inability to tunnel through the Coulomb barrier. With increasing particle energy the tunneling probability, and hence the reaction cross section, increases. The calculation of the tunneling probability for charged particles incident upon a nucleus follows along the lines of the development in Chapter 8 for α-decay. It is customary to parameterize the cross section due to coulombic interactions at low energies as

$$\sigma(E) = \frac{S(E)}{E} e^{-G} \tag{11.28}$$

where the Breit-Wigner equation is written as $S(E)/E$ and is modified by the e^{-G} factor to account for the tunneling probability. Here G is proportional to $E^{-1/2}$ as given in equation (8.17) with Q replaced by the incident particle energy E. At low energies $S(E)$ is more or less constant or varies slowly (as long as no resonance reactions are present in the energy range) and (11.28) may be written as

$$\sigma(E) = \frac{S(0)}{E} e^{-\alpha/\sqrt{E}} \tag{11.29}$$

where the parameter α is determined from equation (8.16) as a function of the nuclear and particle masses, charges, and dimensions. The net result is, that for a given process, there is a threshold energy below which the reaction cross section is virtually zero because the tunneling probability is virtually zero. The resonant peaks corresponding to the population of certain states appear at higher energies superimposed on a background given by equation (11.29). An example of this kind of behavior for a charged incident particle (α-particle) is shown in Figure 11.10.

Figure 11.10   Cross Section for the Reaction $^{13}C(\alpha, n)^{16}O$

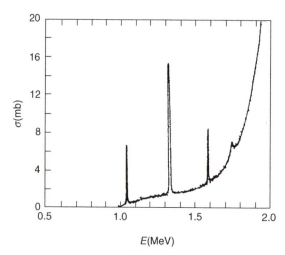

Data are from J. K. Bair and F. X. Haas, *Phys. Rev. C7* (1973), 1356. Copyright 1973 by the American Physical Society.

Problems

11.1. Derive equation (11.14) in the text.

11.2. (a) An α-particle is incident on a ^{13}C nucleus leading to the reaction $\alpha +$ $^{13}C \rightarrow {}^{17}O^*$. Show that this is immediately followed by the decay $^{17}O^* \rightarrow$ $^{16}O + n$.

(b) The cross section for the $^{13}C(\alpha, n)^{16}O$ reaction is shown in Figure 11.10. Determine the energy of the ^{17}O excited state corresponding to the highest energy peak in the figure.

11.3. For a tritium (^3H) nucleus incident on a ^{12}C nucleus construct a graph in the style of Figure 11.4.

11.4. (a) Determine the missing particle/nucleus to complete the following reactions:

   (i) $^{29}Si(\alpha, n)X$
   (ii) $^{60}Ni(x, n)^{60}Cu$
   (iii) $^{111}Cd(n, x)^{112}Cd$
   (iv) $X(p, d)^{188}Os$
   (v) $^{156}Gd(d, x)^{157}Tb$

(b) For each reaction calculate the value of Q.

11.5. Compare the (p, n) reaction to β^--decay. How are the Q values for these two processes related?

11.6. The (d, p) reaction is used to study the excited states of ^{29}Si.

(a) What target nuclei are used in this experiment?

(b) What is the Q for this reaction?

(c) For an incident deuteron energy of 10 MeV and a scattering angle of 30°, what is the final proton energy that corresponds to the ground and first three excited states of ^{29}Si?

11.7. Protons with an initial energy of 8 MeV are scattered elastically from a variety of different nuclei. For a scattering angle of 90°, what is the scattered proton energy when the target nucleus is

   (a) ^7Li, (b) ^{25}Mg, (c) ^{95}Mo, and (d) ^{208}Pb?

11.8. The 4.56 MeV state of ^{17}O is populated by the (n, γ) reaction with ^{16}O (see Figure 11.5). Calculate the excess line width introduced by Doppler broadening at room temperature. Compare with the information for the (n, γ) reaction with ^{238}U shown in Figure 11.9.

12 CHAPTER | **Fission Reactions**

12.1 BASIC PROPERTIES OF FISSION PROCESSES

Fission is the splitting of a relatively heavy nucleus into two lighter nuclei. This can be a spontaneous process or it can be induced by the reaction of the nucleus with an incident particle (usually a neutron). Alpha decay is an extreme case of spontaneous fission where one of the nuclei is very light. Fission usually refers to the situation where the two fragments are of more similar mass. A simple fission process can be described as

$$^{A}X \rightarrow {}^{B}Y + {}^{A-B}Z. \tag{12.1}$$

An investigation of Figure 4.1 shows that if a heavy nucleus breaks up into two lighter nuclei then excess binding energy will be released, that is, the process will be exothermic. Typically a nucleus with (say) $A = 236$ has a binding energy of about 7.6 MeV per nucleon or a total binding energy of 1.78 GeV. If this breaks up into two equal fragments then the two nuclei, with $A = 117$, have binding energies of 8.5 MeV per nucleon or a total of 1.99 GeV. This gives a net energy release of about 210 MeV per fission. The semiempirical mass formula (4.10) can be used to estimate the net energy release as a function of the size of the fission fragments as shown in Figure 12.1. Here the net energy release as a function of B in equation (12.1) is plotted for $A = 236$. This indicates that the energy release is greatest for fission fragments of equal size. The semiempirical mass formula can also be used to determine the energy release for symmetric fission as a function of A in equation 12.1. This is illustrated in Figure 12.2. Clearly this

Figure 12.1 | Energy Release from the Fission of ^{236}U as a Function of A of One of the Fragments

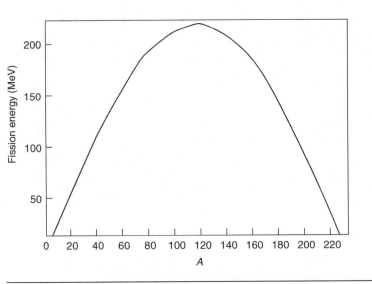

Figure 12.2 | Energy Release from Symmetric Fission as a Function of A of the Parent Nucleus

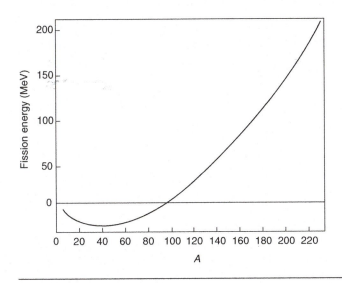

is consistent with the binding energy shown in Figure 4.1, since, if the binding energy per nucleon for nuclei with $A/2$ nucleons becomes less than that for nuclei with A nucleons, then the process will not produce excess energy and is not energetically favorable.

The problem of spontaneous fission is much like that of α-decay because the charged fragments must overcome a coulombic barrier before they can separate. The calculation of the barrier height is not straightforward but can be accomplished by means of the semiempirical mass formula. The results of such a calculation are shown in Figure 12.3. A more detailed calculation involving shell effects gives a similar curve with some additional structure near magic nuclei as shown in the figure by the broken line. These results show that the barrier height drops to zero for A around 300, indicating that nuclei with a mass greater than this are unstable and will undergo spontaneous fission with a very short lifetime. This instability can be understood using the following model.

Figure 12.3 | Coulomb Barrier Energy for Symmetric Fission as a Function of A of the Parent as Calculated from the Semiempirical Mass Formula (solid curve)

The dotted curve illustrates the effect of the nuclear shell structure.

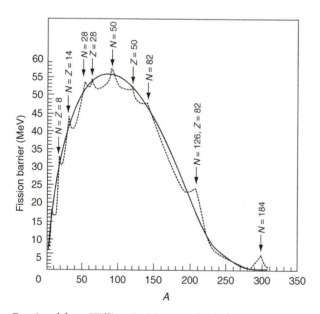

Reprinted from William D. Myers and Wladyslaw J. Swiatecki, "Nuclear masses and deformations," *Nuclear Physics 81* (1966), 1–60, Copyright 1966 with permission from Elsevier Science.

Figure 12.4 | Schematic Illustration of Nuclear Shapes During the Fission Process

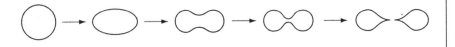

In the context of the liquid drop model, fission can be viewed as the fragmentation of a drop as illustrated in Figure 12.4. Any oscillations in the shape of the drop will tend to grow and to cause the drop to fragment, if the deformation from spherical symmetry makes the drop more stable. We can calculate the energy associated with nuclear deformation (or stretching as shown in the figure) in terms of nuclear binding energies as

$$\Delta E = B(\varepsilon) - B(\varepsilon = 0) \tag{12.2}$$

where ε is a deformation parameter. If it is assumed that the deformations are elliptical, then ε is the eccentricity. Since nuclear matter is not compressible the volume of the nucleus does not change during deformation. We can, therefore, equate the volume of the spherical nucleus of radius R_0 before stretching to the volume of the stretched ellipsoidal nucleus as

$$\frac{4\pi}{3} R_0^{\,3} = \frac{4\pi}{3} ab^2 \tag{12.3}$$

where a and b are the semimajor and semiminor axes of the ellipsoid, respectively. The eccentricity is then related to a and b as

$$a = R_0(1 + \varepsilon)$$
$$b = R_0(1 + \varepsilon)^{-1/2}. \tag{12.4}$$

Using the semiempirical mass formula we can now calculate the energy associated with the deformation. We anticipate that changes in the binding energy will be manifested in the surface and Coulomb terms of the semiempirical mass formula. The surface area of an ellipsoid can be expressed as

$$S = 4\pi R_0^{\,2}\left(1 + \frac{2}{5}\varepsilon^2 + \cdots\right) \tag{12.5}$$

where R_0 is related to ε by equation (12.4). The Coulomb energy of an ellipsoid relative to that of a sphere of the same volume can be expressed as

$$\frac{E_{ellipse}}{E_{sphere}} = 1 - \frac{1}{5}\varepsilon^2 + \cdots. \tag{12.6}$$

Taking terms to order ε^2, the semiempirical mass formula gives ΔE from equation (12.2) as

$$\Delta E = -a_S A^{2/3}\left(1+\frac{2}{5}\varepsilon^2\right) - a_C Z(Z-1)A^{-1/3}\left(1-\frac{1}{5}\varepsilon^2\right) + a_S A^{2/3} + a_C Z(Z-1)A^{-1/3}$$

(12.7)

or

$$\Delta E = \left[-\frac{2}{5}a_S A^{2/3} + \frac{1}{5}a_C Z(Z-1)A^{-1/3}\right]\varepsilon^2.$$

(12.8)

ΔE will be positive and the nucleus will be unstable to stretching if

$$\frac{Z(Z-1)}{A} > \frac{2a_S}{a_C}.$$

(12.9)

Using values of the semiempirical mass formula parameters from Chapter 4 gives

$$\frac{Z(Z-1)}{A} > 50.$$

(12.10)

Heavy nuclei have $Z/A = 0.4$ and equation (12.10) gives $A = 300$ as the maximum mass of fission stable nuclei, consistent with the barrier energy as shown in Figure 12.3.

12.2 INDUCED FISSION

In some nuclei, fission can be induced by bombardment with neutrons. Uranium provides an interesting example as it is commonly used for fuel in commercial fission reactors. Natural uranium consists of approximately 0.72% ^{235}U and 99.27% ^{238}U. A low energy neutron can be captured by either of these nuclei yielding the following reactions

$$n + {}^{235}U \rightarrow {}^{236}U + 6.46 \text{ MeV}$$

(12.11)

$$n + {}^{238}U \rightarrow {}^{239}U + 4.78 \text{ MeV}.$$

(12.12)

In heavy nuclei the density of states, especially at energies a few MeV above the ground state, is generally quite high and there will be many excited states available that can be occupied. An inspection of Figure 12.3 shows that the barrier energy for $A = 236$ is about 6.2 MeV. Thus the process in equation (12.11) leaves the ^{236}U nucleus with enough excess energy to induce fission without the need to tunnel through the Coulomb barrier. The process in equation (12.12) does not produce enough energy to induce fission and about 1.4 MeV additional is needed (which can be supplied by additional kinetic energy associated with the neutrons). Nuclides in which fission can be induced by low energy (that is, thermal) neutrons are referred to as *fissile* while those materials in which thermal neutrons will not induce fission because the energy is below the barrier are called *nonfissile*. For heavy nuclides, odd A nuclides are generally fissile while even–even nuclides are nonfissile. This is seen in the example in equations (12.11) and (12.12), and can be understood by a consideration of the pairing term in the semiempirical mass formula.

In a sample of fissile material, a spontaneous fission will produce excess neutrons and these will induce further fissions. Under appropriate conditions a chain reaction can occur. Since, from Figure 3.1, heavier β-stable nuclei have a greater ratio N/Z, the two fragments resulting from the fission of a heavy nucleus will require less than the total number of neutrons available. Thus equation (12.1) may be more correctly written as

$$^A X \rightarrow {}^B Y + {}^{A-B-\nu} Z + \nu n. \qquad (12.13)$$

where, on the average, ν excess neutrons are given off in a fission process. Even considering the excess neutrons given off during the fission process, the fission fragments Y and Z are typically too rich in neutrons and will decay towards the β-stability line by β^--decay. As these nuclides approach the β-stability line, the lifetime becomes longer and this is the source of the long-lived radioactive waste produced by fission reactors.

12.3 FISSION PROCESSES IN URANIUM

In most reactors it is fission of uranium nuclei that supplies the energy and this is primarily a result of the behavior of ^{235}U. The induced fission process in ^{235}U is written as

$$n + {}^{235}U \rightarrow {}^B Y + {}^{236-B-\nu} Z + \nu n \qquad (12.14)$$

where, on the average ν is about 2.5. These excess neutrons are referred to as prompt neutrons and are given off on the time scale of the fission process, about 10^{-14} seconds. The distribution of masses for the fission fragments from ^{235}U (and some other fissile nuclides) is illustrated in Figure 12.5. This illustrates that the fission process does not result in equal-sized fragments. By definition this graph must be approximately symmetric since every fission that produces a fragment that is larger than $A/2$ must also produce a fragment that is smaller than $A/2$ by the same amount. Minor variations result because of the excess neutrons that are emitted.

The fission fragments are normally left in excited states that γ-decay to their ground states. These γ-rays are referred to as prompt γ-rays. The energy that is immediately released by the fission process is distributed between the

Figure 12.5 | Distribution of Values of *A* for the Fragments from the Induced Fission of Some Fissile Nuclei

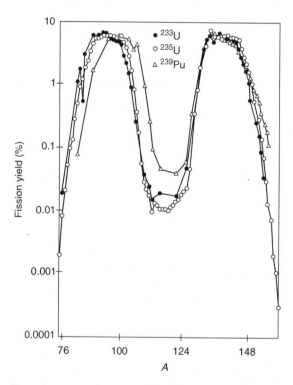

From A. M. Weinberg and E. P. Wigner, *The Physical Theory of Neutron Chain Reactions*. Chicago: University of Chicago Press, 1958. Copyright University of Chicago Press, 1958. Used with permission.

kinetic energy of the fission fragments, the kinetic energy of the prompt neutrons, and the energy of the prompt γ-rays as shown in Table 12.1. Primarily it is this energy that becomes available as heat and that can be extracted from the reactor.

The neutron rich fission fragments decay by a series of β^-- and γ-decays toward β-stability. This results in a delayed release of energy due to electron and antineutrino kinetic energy (the latter of which is lost into space) and γ-ray energy as given in the table. In some cases, a nuclide in the β-decay sequence is left in an excited state that is above the neutron separation threshold and can therefore decay by neutron emission. These are referred to as delayed neutrons and on the average there are about 0.02 neutrons per fission given off on a time scale of tens of seconds after the fission event. Although these represent a negligible amount of energy they are very important in controlling the chain reaction as discussed in the material that follows.

Table 12.1 | Distribution of Energy from the Fission of ^{236}U

Source of Energy	E (MeV)
fission fragment kinetic energy	167
prompt neutron kinetic energy	5
prompt γ-ray energy	6
delayed β-decay energy	8
delayed γ-ray energy	7
delayed antineutrino energy	12
total	205

12.4 NEUTRON CROSS SECTIONS FOR URANIUM

In order to understand the details of induced fission in uranium it is essential to investigate the neutron cross section as a function of energy. There are a number of possible processes by which the neutrons can interact with nuclei. These are discussed below.

Elastic Scattering   It is important to realize that elastic scattering conserves kinetic energy. However, as the target nucleus gains energy when it recoils, the neutron loses some small amount of energy.

Inelastic Scattering   In this process the neutron gives up some of its kinetic energy to the target nucleus leaving it in an excited state. This process has a threshold energy equal to the energy of the first excited state above the ground state of the nucleus. For ^{235}U and ^{238}U, the inelastic scattering threshold is 14 keV and 44 keV, respectively.

Radiative Capture   In the (n, γ) reaction, a neutron is absorbed and the resulting nucleus decays by γ-emission to a state below the neutron separation energy, thereby capturing the neutron. The cross section for (n, γ) process is characterized by a series of resonances as discussed in the last chapter. The lowest energy resonance occurs at an energy equal to the difference between the neutron separation energy and the energy of the next available state.

Fission   In this process the neutron leaves the nucleus in an energy state above the fission barrier and fission proceeds spontaneously on a very short time scale. The energy available is the sum of the neutron kinetic energy and the Q for the process. The energy dependent neutron induced fission cross sections for ^{235}U and ^{238}U are illustrated in Figure 12.6.

Figure 12.6 | Fission Cross Sections for Neutrons Incident on (a) ^{235}U and (b) ^{238}U as a Function of Energy

The figure illustrates the general features but not the details of the cross section for ^{235}U in the region of (n, γ) resonances.

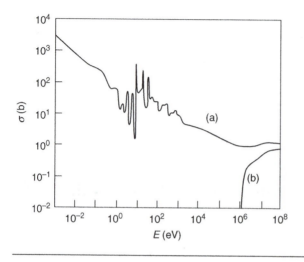

The total neutron cross section is the sum of the four contributions described above. In order to understand the relative importance of these four processes at different energies we divide the energy scale into four regions; $E < 1$ eV, 1 eV $< E < 100$ eV, 100 eV $< E < 1$ MeV, and $E > 1$ MeV, and discuss the behavior of ^{235}U and ^{238}U for each of these regions.

Cross Sections for ^{235}U

$E < 1$ eV   The cross section is dominated by the fission cross section and this accounts for about 85% of the total. The remaining cross section is predominantly due to (n, γ) processes as the result of a ^{236}U resonance lying just below $E = 0$.

1 eV $< E < 100$ eV   This region is dominated by the (n, γ) resonances with the remaining cross section being fission and to a lesser extent, elastic scattering. It is important to realize that scattering processes reduce the energy of the neutrons while (n, γ) processes absorb them.

100 eV $< E < 1$ MeV   (n, γ) processes are important here as well but the energy levels of the excited states overlap considerably and the resonances no longer show discrete peaks. Above the 14 keV threshold, inelastic scattering processes are important. Fission processes become less important as the energy increases.

E > 1 MeV   Processes are similar to those in the previous energy range but with a decreasing probability of (n, γ) processes.

Cross Sections for ²³⁸U

E < 1 eV   Elastic scattering is the only possible process in this region.

1 eV < E < 100 eV   Resonances due to (n, γ) processes and some elastic scattering; see Figure 11.8.

100 eV < E < 1 MeV   Unresolved (n, γ) resonances, some elastic scattering and inelastic scattering above the threshold energy of 44 keV.

E > 1 MeV   Predominantly inelastic scattering and a reduced probability for (n, γ) reactions. Fission above a threshold energy of about 1.4 MeV.

Neutrons that are given off during the fission process in ²³⁵U or ²³⁸U have a kinetic energy of about 2 MeV. Based on the information above we can understand the interaction of these neutrons with uranium nuclei. A simple example is given in the next section.

12.5 CRITICAL MASS FOR CHAIN REACTIONS

In a sample of uranium a chain reaction is produced if the $(v-1)$ excess neutrons produced by a fission event induce more than one additional fission. In a controlled chain reaction the neutrons produced by one fission will induce exactly one more fission. The remaining neutrons will be lost either by (n, γ) processes or by exiting the sample. In order to understand this problem quantitatively let's look at an example of a piece of pure ²³⁵U. If each fission neutron will, on the average, produce vq neutrons then there will be a net gain of $(vq - 1)$ neutrons on the time scale of the fission, τ. The value of $q < 1$ and accounts for the neutrons that are lost. Thus if $n(t)$ neutrons are present at time t then at time $t + dt$ there will be

$$n(t + dt) = n(t)\left[1 + (vq - 1)\frac{dt}{\tau}\right] \qquad (12.15)$$

neutrons. This can be written as

$$\frac{dn}{dt} = \frac{(vq - 1)n(t)}{\tau} \qquad (12.16)$$

and has the solution

$$n(t) = n(0)\exp\left[\frac{(vq - 1)t}{\tau}\right]. \qquad (12.17)$$

If $vq - 1$ is negative then $n(t)$ becomes small and the chain reaction dies out, if $vq - 1$ is positive then $n(t)$ becomes very large very fast and the chain reaction is uncontrolled. It is only when $vq = 1$ that the number of neutrons remains constant and the chain reaction is controlled. Since $v = 2.5$ this implies a value of q about 0.4 for a controlled chain reaction. In order to see how vq can be controlled we need to determine how large τ is and how far the neutrons travel during that time. From Figure 12.6 we see that a 2 MeV neutron in ^{235}U has a fission cross section of about 1 barn. This may be compared with the total neutron cross section for ^{235}U at this energy, which is about 7 barns (the nonfission part being primarily inelastic scattering). Thus, on the average, a neutron will undergo one fission per seven interactions. The distance that the neutron travels between interactions (that is, the mean free path) is

$$l = \frac{1}{\rho\sigma} \tag{12.18}$$

where ρ is the number density of nuclei in the material and σ is the total cross section. For uranium $\rho = 4.8 \times 10^{28}$ m^{-3} and, using $\sigma = 7$ barns, we find a mean free path of 0.03 m. A simple nonrelativistic calculation shows that a 2 MeV neutron will traverse this distance in 1.5×10^{-9} sec. Because, on the average, seven interactions are necessary to induce one fission, then the total time between fissions is $7 \times 1.5 \times 10^{-9} = 10^{-8}$ sec (since the actual fission process is much faster). The total distance traveled is found by considering the neutron's path as a random walk, because each inelastic scattering event will change the direction of the neutron's path in a random way. The total distance traveled from the origin is given as $\sqrt{7} \times 0.03$ m $= 0.079$ m. Since the (n, γ) reaction is not a significant contribution to neutron loss in this energy region, then a sphere of ^{235}U with a radius much less that 0.079 m will lose most of the fission neutrons before they induce further fission events, and the chain reaction will die out. A sphere of radius much more than 0.079 m will have q close to unity and the reaction will be uncontrolled. The size of the ^{235}U sample at which the chain reaction is sustained is called the critical radius (and the mass is called the critical mass). A more detailed calculation for ^{235}U gives a critical radius of 0.087 m corresponding to a critical mass of about 52 kg.

12.6 MODERATORS AND REACTOR CONTROL

In a natural mixture of ^{235}U and ^{238}U there is a small probability of fission being induced by 2 MeV neutrons. Since the concentration of ^{235}U is small (0.72%) the probability of ^{235}U fission is also small. The majority of neutrons will scatter inelastically with ^{238}U nuclei until their energy is below the fission threshold of 1.4 MeV. The greatest probability of inducing fission is at low energy in ^{235}U. Even though the fraction of ^{235}U is small the fission cross section is three orders of magnitude or more greater than at high energy for either ^{235}U or ^{238}U. In a reactor using natural uranium or uranium only slightly enriched in ^{235}U there are two major design problems to overcome: how to reduce the energy of the high energy neutrons to the range below about 1 eV without losing neutrons to the (n, γ) process (primarily in the ^{238}U) and how to sustain the chain reaction in a controlled way.

Figure 12.7 | Basic Design of the Core of a Fission Reactor

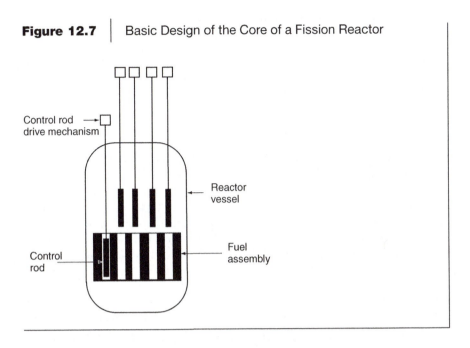

The most common design of a nuclear reactor is the thermal reactor that uses natural or only slightly ^{235}U enriched uranium as a fuel. This reactor design utilizes a core consisting of number of uranium fuel rods, each of which is less than the critical mass (typically about 1 cm in diameter). The rods are bundled together to form fuel assemblies in arrays of typically 8×8 to 16×16 rods (depending on the reactor design). The fuel assemblies are surrounded by a moderator that is contained in a reactor vessel and are separated by control rods as illustrated in Figure 12.7. Fast neutrons within a fuel rod have a small probability of inducing fission within the rod but will most likely be emitted into the moderator while their energy is still in the MeV range. The moderator is a material that effectively decreases the energy to a fraction of an eV. These thermalized neutrons are then incident upon another fuel rod where they will most likely induce a fission process in ^{235}U because of the large fission cross section in this isotope at low energies (see Figure 12.6). The control rods are made of a material that will effectively stop the neutrons and are used to limit the number of neutrons that travel between fuel rods allowing the value of q in equation (12.17) to be controlled. We briefly discuss the criteria for suitable materials for the moderator and control rods below.

Moderator   The purpose of the moderator is to allow the neutron energy to be reduced in the region outside the fuel rod, thus avoiding neutron loss from (n, γ) reactions in the uranium. The moderator is a material that interacts with neutrons over a wide range of energies by elastic and/or inelastic scattering, thus reducing the neutron energy without reducing the number of neutrons. Desirable criteria for the choice of moderator are as follows:

1. Inexpensive and easily obtained
2. Small (n, γ) cross section over a wide range of energies
3. Small nuclear mass, in order to maximize the energy transfer per scattering event
4. Reasonably high density
5. Chemically stable with a low level of toxicity

Various materials satisfy some of these criteria although no material is ideal. Graphite satisfies all criteria except number 3 requiring that a larger quantity of moderator material is used. Water satisfies all criteria except number 2 and requires the use of ^{235}U enriched fuel. Heavy water (D_2O) has a lower (n, γ) cross section than (H_2O) but the processes that do occur $(n + d \rightarrow t + \gamma)$ produce radioactive tritium as an undesirable byproduct. In general, the properties of the moderator determine the quantity of moderator needed as well as the degree to which the uranium has to be enriched in ^{235}U for the reactor to operate efficiently.

Control Rods   The control rods should have a large neutron absorption (that is, (n, γ)) cross section, especially at low energies. Cadmium metal is the most commonly used material and has a low energy neutron cross section as shown in Figure 12.8. The large peak at a fraction of an eV is due to an (n, γ) process in ^{113}Cd that constitutes 12.3% of natural Cd. The low energy $1/E$ dependence as given in equation (11.23) is seen in the figure.

Figure 12.8 | Total Neutron Cross Section for Natural Cd

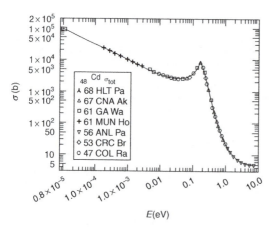

From D. I. Garber and R. R. Kinsey, *Neutron Cross Sections, Vol. II.* Upton, N.Y.: Brookhaven National Laboratory, 1976. Courtesy of Brookhaven National Laboratory.

12.7 REACTOR STABILITY

In principle it should seem that it would merely be a matter of regulating the position of the control rods between the fuel elements in order to control the number of neutrons that can induce fission and thereby adjust the value of vq in equation (12.17). The problem here, however, is a matter of the time scale involved. As we have seen the average lifetime of a fission neutron in ^{235}U is about 10^{-8} seconds. In a fission reactor the neutrons spend much of their time in the moderator and their lifetime can be several orders of magnitude longer than given by our simple calculation. However, this time scale (perhaps 10^{-4} sec) is still too short to mechanically adjust the position of control rods in the reactor. The key to reactor control lies with the delayed neutrons.

Each fission process produces v' delayed neutrons (about 0.02) in addition to the v prompt neutrons. The critical condition as described above is now

$$(v + v')q = 1. \tag{12.19}$$

If the reactor is designed in such a way that $vq = 0.99$ then the control rods can be utilized on a time scale of tens of seconds to regulate the delayed neutrons to insure that condition (12.19) is met.

An important factor in reactor stability is the influence of temperature on the parameter q. As a result of energy release due to fission the reactor components naturally increase in temperature. For safety reasons it is important that

$$\frac{dq}{dT} < 0. \tag{12.20}$$

If the opposite were true then temperature increases would lead to an increased q and could result in an uncontrolled chain reaction. One of the major factors influencing dq/dT is the Doppler broadening of the (n, γ) resonances in the fuel rods. Although most of the neutrons that are lost are absorbed in the control rods, a small number pass through a fuel rod when their energy is in the range of the ^{238}U (n, γ) resonances as seen in Figure 11.8 and are lost by radiative capture. As the temperature increases and these resonances broaden (see Section 11.5) the probability that a neutron will have the correct energy to be absorbed increases. This additional probability of neutron absorption corresponds to a decrease in q as temperature increases.

12.8 REACTOR DESIGN

There are numerous designs of nuclear reactors. The simplest design is perhaps the boiling water reactor. In this design the moderator (H_2O) is allowed to boil and the steam is used to run a turbine that in turn drives an electric generator. This design has the disadvantage that circulating the moderator outside the containment vessel could possibly lead to the spread of radioactive contamination. A preferable design is illustrated in Figure 12.9 where pressurized H_2O in a closed system is used as the moderator and transfers heat through a heat

Figure 12.9 | Diagram of a Fission Reactor Using Pressurized H_2O as a Moderator

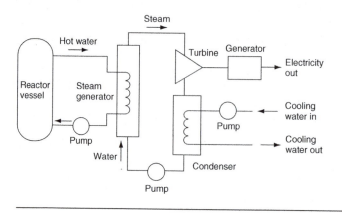

exchanger to a steam generator. The properties of H_2O as a moderator require that the uranium in these reactors be enriched to 2–3% ^{235}U. This type of design is the most common fission reactor in commercial use in the United States.

A similar design in common use in Canada (the CANDU reactor) uses D_2O in a closed system as a moderator. This reactor has the advantage that it can use natural uranium as a fuel. In both these designs the H_2O or D_2O serves both as a moderator and to transfer heat to a steam generator. This latter function is necessary not only to extract energy from the reactor but also to cool the reactor core and prevent it from overheating. Graphite moderator reactors are in common use in Great Britain. These typically use a closed system of circulating He gas to cool the core and transfer heat to a heat exchanger.

An obvious drawback in the design of a fission reactor as described above is that it makes use of the fission energy from ^{235}U but does not utilize the available energy from ^{238}U. Thus only about 1% of the potential energy from natural uranium is extracted. A solution to this problem is the breeder reactor. Excess fission neutrons are used to generate fissile fuel from nonfissile ^{238}U. Neutron capture produces ^{239}U that in turn decays by β-decay:

$$^{239}U \rightarrow {}^{239}Np + e^- + \bar{\nu}_e. \tag{12.21}$$

This is followed by

$$^{239}Np \rightarrow {}^{239}Pu + e^- + \bar{\nu}_e. \tag{12.22}$$

The lifetime is about 40 minutes for the first β-decay and about four days for the second. Fissile ^{239}Pu is β-stable and can be readily produced by the above reactions. This has fission properties that are very similar to those of ^{235}U and is suitable for reactor fuel use.

As a final note concerning fission reactors from a commercial standpoint a historical perspective is of interest. Throughout the 1960s and 1970s numerous reactors were constructed worldwide and many are still functional today. In the late 1970s reactor construction in the United States stopped and no new reactors have been built since then. Safety concerns were, at least partially, responsible for the waning interest in nuclear fission reactors. In recent years there has been renewed interest in fission as an energy source and a number of new reactors are under construction or are in the planning stages outside of North America. This new interest is not in breeder reactors, because prototype breeders do not have a proven safety record, but for conventional enriched uranium reactors. Improved safety is a major factor in the new reactor designs. Many utilize very small fuel pellets that are less than a mm in diameter and are coated with ceramic (to prevent spread of radioactive material in the event of overheating) and are embedded in a graphite matrix (which acts as the moderator). Cooling and heat transfer is by circulating helium gas, similar to the British reactor design. The viability of this technology from scientific, economic, and safety standpoints remains to be seen, as does the future of fission energy throughout the world.

Problems

12.1. Figure 11.8 shows the neutron absorption cross section for ^{238}U as a function of the center of mass energy. Calculate the ^{239}U excited state energies corresponding to the peaks shown in the figure. Comment on the results of this calculation.

12.2. (a) A neutron with initial kinetic energy T_i collides elastically with a stationary nucleus of mass M. On the average the neutron energy after the collision may be written as

$$T_f = (M^2 + m_n^2)T_i/(M + m_n)^2.$$

For a Uranium fission neutron travelling in a water moderator, estimate the number of elastic collisions necessary to reduce energy from 2.0 MeV to 0.01 eV. Assume that the elastic scattering cross section for neutrons on ^1H is 0.33 b and is independent of energy below 2.0 MeV. Assume the cross section for ^{16}O is negligible.

(b) Estimate the time required for the reduction of neutron energy in part (a).

12.3. Use the semiempirical mass formula to derive an expression for the curve shown in Figure 12.1.

12.4. Use the semiempirical mass formula to derive an expression for the curve shown in Figure 12.2.

12.5. A fission reactor produces 3×10^9 W of electrical power. Assuming that the steam generator has an efficiency of 30%, calculate the mass of ^{235}U consumed by the reactor in one day.

12.6. ^{265}Fm decays almost exclusively by spontaneous fission with a lifetime of 3.9 h. Assuming that Fm has a molar specific heat of 25 J/(mole · K) and assuming that all fission energy is converted into thermal energy within the Fm, estimate the increase in the internal temperature of a 1 μg sample of Fm due to fission during 1 second.

12.7. Consider a sample of ^{235}U with a mass of 1000 kg, that is, much greater than the critical mass so that $q \approx 1$. At $t = 0$ a single spontaneous fission neutron initiates a chain reaction. Calculate the time required for all ^{235}U nuclei in the sample to be consumed in fission reactions.

12.8. Determine which isotopes of Pu with lifetimes of greater than a day are fissile.

Fusion Reactions

13.1 FUSION PROCESSES

We saw in the last chapter how energy can be extracted by breaking up heavy nuclei into lighter nuclei and thereby making use of the increase in binding energy per nucleon. Along similar lines, we can see from Figure 4.1 that combining two light nuclei to make a heavier nucleus (up to $A = 55$) also yields an increase in binding energy per nucleon and this process may be used as a source of energy. This process is referred to as fusion and is the process by which the sun and other stars produce energy. Fusion is a desirable method of producing energy and has several significant advantages over fission. These include

1. An inexpensive and plentiful supply of fuel
2. Reactions that are inherently easier to control and are therefore much safer
3. Substantially reduced environmental hazards from reactor byproducts

Unfortunately, at present, fusion power is not technologically feasible. This is because of the fundamental differences between induced fission and the fusion process. For a fissile material, fission is induced by a thermal neutron because the resulting nuclear state is above the Coulomb barrier. For fusion, the Coulomb barrier, which is in the range of a few MeV or a few tens of MeV, is always a consideration. It is certainly straightforward to accelerate nuclei to these energies in even very small particle accelerators and to collide them with other nuclei to produce fusion reactions. Unfortunately, in such a situation, the energy expenditure is substantially greater than the energy gain. Thus, although this is a useful way of learning about fusion reactions in the laboratory, fusion, in the

175

practical sense, always involves energies below the Coulomb barrier and inevitably involves a tunneling process. A treatment of this phenomenon is analogous to that for α-decay where the barrier height can be expressed as

$$V_C = \frac{e^2}{4\pi\varepsilon_0} \frac{Z_1 Z_2}{R_1 + R_2}. \tag{13.1}$$

Here Z_i and R_i are the charges and radii of the two nuclei, respectively. In order to maximize the tunneling probability it is necessary to minimize the barrier height. From equation (13.1) it is, therefore, obvious that nuclei with small values of Z are the most interesting. These are involved in the processes that are primarily responsible for energy production in the sun and also those that have attracted interest as possible sources of fusion power. In this section we consider some of the possible fusion reactions involving isotopes of hydrogen.

The simplest fusion process might appear to be the fusion of two protons, the so-called p–p process. However, two protons cannot form a bound state and p–p fusion is analogous to β^+-decay where one of the protons is converted to a neutron to give

$$p + p \rightarrow {}^2H + e^+ + \nu_e. \tag{13.2}$$

The energy release from this process is 0.42 MeV per fusion. Normally an additional 1.02 MeV will become available due to the annihilation of the positron with an electron. Like the β-decay processes discussed previously, the presence of leptons in the reaction indicates that the weak interaction is responsible. As a result the cross section for this reaction is very small.

Fusion processes involving deuterons are of importance for our further discussions. The simplest of these is

$$d + p \rightarrow {}^3He + \gamma \qquad (Q = 5.49 \text{ MeV}). \tag{13.3}$$

We will see in Section 13.3 that this is an important reaction in stars. The most obvious process involving the fusion of two deuterons is the formation of ${}^4He$:

$$d + d \rightarrow {}^4He + \gamma \qquad (Q = 23.8 \text{ MeV}). \tag{13.4}$$

This process is unlikely, as the energy release is well above the neutron and proton separation energies of ${}^4He$. Consequently there are two possible modes of d–d fusion:

$$d + d \rightarrow {}^3H + p \qquad (Q = 4.0 \text{ MeV}) \tag{13.5}$$

and

$$d + d \rightarrow {}^3He + n \qquad (Q = 3.3 \text{ MeV}). \tag{13.6}$$

A final process of importance is the fusion of a deuteron with a triton (d–t fusion):

$$d + t \rightarrow {}^4He + n \qquad (Q = 17.6 \text{ MeV}). \qquad (13.7)$$

This process releases a substantial amount of energy and is of particular importance, as we will see in Section 13.4, for the operation of a controlled fusion reactor.

It is important to understand how the fusion energy is distributed among the fusion byproducts. In the case of a process, such as given in equation (13.7), where two relatively massive particles appear after the fusion process, a simple consideration of energy and momentum conservation is useful. For a generic reaction X(a, b)Y the Q of the reaction is given by

$$Q = [m_X + m_a - m_Y - m_b]c^2. \qquad (13.8)$$

The kinetic energy of the fusion byproducts is therefore,

$$Q = \frac{1}{2}m_Y v_Y^2 + \frac{1}{2}m_b v_b^2 \qquad (13.9)$$

where we have assumed that the incident kinetic energy is small. This is a reasonable assumption for problems dealing with stellar processes and controlled fusion reactors. Conservation of momentum gives,

$$m_Y v_Y = m_b v_b. \qquad (13.10)$$

The above equations can be solved to give the energy of the two particles as

$$E_Y = \frac{Q}{1 + \dfrac{m_Y}{m_b}} \qquad (13.11)$$

and

$$E_b = \frac{Q}{1 + \dfrac{m_b}{m_Y}}. \qquad (13.12)$$

From these expressions the ratio of kinetic energies is given by

$$\frac{E_b}{E_Y} = \frac{m_Y}{m_b}. \qquad (13.13)$$

It is readily seen that the less massive particle acquires the larger fraction of the energy. This is particularly important, for example, in the case of d–t fusion where 80% of the energy is carried away by the neutron. Because of the neutron's low reaction cross section, the method by which this energy may be extracted requires careful consideration.

13.2 FUSION CROSS SECTIONS AND REACTION RATES

The cross section for fusion follows from the cross section for charged particle reactions as given by equation (11.28):

$$\sigma \propto \frac{1}{v^2} e^{-G}. \tag{13.14}$$

Some numerical factors have been omitted here but the expression above contains all energy or velocity dependencies. The factor G in the exponent is determined by the probability of tunneling through the Coulomb barrier between the two nuclei. This follows directly from the discussion of α-decay and equation (8.17) as:

$$G = \frac{2Z_1 Z_2 e^2}{4\pi\varepsilon_0 \hbar} \sqrt{\frac{2m}{E}} \left[\cos^{-1}\sqrt{\frac{a}{b}} - \sqrt{\frac{a}{b}\left(1 - \frac{a}{b}\right)} \right] \tag{13.15}$$

where m is the reduced mass of the system. Along the lines of the discussion in Chapter 8, the quantities a and b are defined as

$$a = R_1 + R_2 \tag{13.16}$$

and

$$b = \frac{Z_1 Z_2 e^2}{4\pi\varepsilon_0 E}. \tag{13.17}$$

Since the kinetic energy of the nuclei involved is much less than the Coulomb barrier energy, the term in brackets in equation (13.15) can be approximated as $\pi/2$. Using the center of mass kinetic energy, $E = mv^2/2$, the cross section for fusion is given by equation (13.14) with

$$G = \frac{2Z_1 Z_2 e^2}{4\pi\varepsilon_0} \frac{\pi}{\hbar v}. \tag{13.18}$$

Some examples of energy dependent cross sections of important fusion reactions are shown in Figure 13.1. The reaction rate for fusion can be determined by considering a beam of particles of species "1" with a flux Φ_1 incident on a collection of particles of species "2" with a number density n_2 and a relative reaction cross section σ. In this case, the reaction rate per unit volume will be given by

$$R = \sigma \Phi_1 n_2. \tag{13.19}$$

For a plasma consisting of two types of particles, as would be appropriate for the interior of a star or the fuel of a fusion reactor, the flux in equation (13.19) is written in terms of n_1, the number density of particles of species "1," and the relative particle velocity, v, as

$$\Phi_1 = n_1 v. \tag{13.20}$$

Figure 13.1 | Energy Dependence of the Reaction Cross Section for d–d Fusion (Total of Equations (13.5) and (13.6)) and d–t Fusion (Equation (13.7))

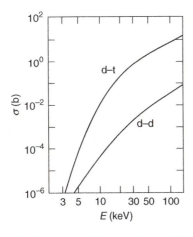

This gives a reaction rate of

$$R = \frac{n_1 n_2 \langle \sigma v \rangle}{1 + \delta_{12}} \tag{13.21}$$

where the delta function in the denominator is "0" for the case where the two species of particles are different and "1" for the case where they are the same. This avoids double counting for a plasma consisting of a single type of nuclei. Since both the cross section and the velocity are functions of the particle energy, it is appropriate to determine the quantity $\langle \sigma v \rangle$ in equation (13.21) in terms of the energy distribution of the particles in the plasma. The Maxwell-Boltzmann distribution gives the probability of finding a particle with a velocity between v and $v + dv$ at a temperature T as

$$P(v)dv = \left(\frac{2}{\pi}\right)^{1/2} \left(\frac{m}{k_B T}\right)^{3/2} v^2 e^{-E/k_B T} dv. \tag{13.22}$$

This gives the average value of $\langle \sigma v \rangle$ as

$$\langle \sigma v \rangle = \int \sigma v P(v) dv \tag{13.23}$$

or

$$\langle \sigma v \rangle \propto \int e^{-G} e^{-E/k_B T} dE. \tag{13.24}$$

Thus the reaction rate is determined by the product of the energy dependent cross section and the Maxwell-Boltzmann distribution. An example of combining

Figure 13.2 | Effect of Combining the Maxwell-Boltzmann Energy Distribution, dN/dE, and the Fusion Cross Section, $\sigma(E)$, to Give the Overall Reaction Rate, $\sigma dN/dE$

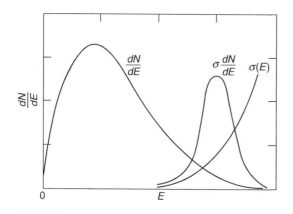

these two factors is illustrated in Figure 13.2. This indicates that there is a maximum in the reaction rate as a result of the decreasing cross section at low energies and the decreasing Maxwell-Boltzmann distribution of high energies. In most cases the energies involved in stellar fusion and fusion reactors are relatively low and it is the portion of the reaction rate curve that increases with energy that is of importance. This is illustrated in Figure 13.3 where the product

Figure 13.3 | Energy Dependence of $\langle \sigma v \rangle$ for d–d (Δ), d–t (O), and Other Fusion Processes

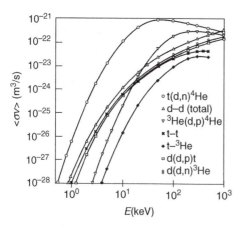

From D. Keefe, *Ann. Rev. Nucl. Part. Sci.* 32 (1982), 391.

$\langle \sigma v \rangle$ is given for d–d, d–t, and other fusion reactions over the range of energies that is most relevant. The peak reaction rate for d–t fusion occurs at around 30 keV, while the peak for d–d fusion occurs well into the MeV range and is not seen in the graph.

13.3 STELLAR FUSION PROCESSES

The sun, like most other stars, produces energy by fusing four hydrogen nuclei (^{1}H) into one helium nucleus (^{4}He). Fusion processes involving heavier nuclei are important in some stars but are relatively unimportant in the sun for two reasons:

1. The concentration of heavier nuclei is relatively small (the sun is about 92% hydrogen and about 8% helium, in terms of the number of atoms).
2. Reaction rates for heavier nuclei are small because of the larger Coulomb barriers involved.

The fusion of four ^{1}H nuclei represents the process

$$4p \rightarrow {}^4He + 2e^+ + 2\nu_e \qquad (13.25)$$

where two of the fusing protons must be converted into neutrons by β^+-decay processes. It is important to note that this is a nuclear process and in order to deal with the energetics in terms of atomic masses it is necessary to add four electrons to each side to give

$$4\,{}^1H \rightarrow {}^4He + 2e^- + 2e^+ + 2\nu_e. \qquad (13.26)$$

The four hydrogen nuclei do not fuse simultaneously to form helium. Instead, the helium forms in a series of steps. The first step of this fusion process is the fusion of two protons as given by equation (13.2). In principle, two deuterons could then fuse according to equation (13.4) to form a ^{4}He nucleus. However, the low concentration of deuterons in the sun as well as the large value of Q for this process, as discussed previously, make this process highly unlikely. A more likely process is p–d fusion as given by equation (13.3) to form ^{3}He. Because of the high concentration of protons, the most logical process involving ^{3}He would seem to be the formation of ^{4}Li. However, ^{4}Li (three protons and one neutron) does not form a bound state and immediately leads to the process

$$^3He + p \rightarrow {}^4Li \rightarrow {}^3He + p. \qquad (13.27)$$

The reaction of ^{3}He with a deuteron is unlikely as the deuterons that are formed relatively quickly fuse with protons to form more ^{3}He. Thus, a ^{3}He nucleus's most likely fate will be to eventually react with another ^{3}He nucleus to form ^{4}He according to the reaction

$$^3\text{He} + {}^3\text{He} \rightarrow {}^4\text{He} + 2{}^1\text{H} + \gamma \qquad (Q = 12.86 \text{ MeV}). \qquad (13.28)$$

The overall process described above is the most common method of energy production in the sun and is referred to as the proton–proton cycle. It has the net result of fusing four protons and in conjunction with two β-decay processes forms a ^{4}He nucleus. The total energy associated with this process is $Q = 26.7$ MeV. Most of this energy is eventually converted into solar radiation; a small amount is carried away as kinetic energy by the neutrinos from the β-decay processes and is lost. The rate at which this process can proceed is limited by the weak interaction cross section for deuteron production as indicated in equation (13.2).

Other reactions that ultimately result in the fusing of four hydrogen into one helium are also possible in the sun and these are described in Figure 13.4. The branching ratios for the various reactions are indicated in the figure. Details of some of these reactions will be discussed in Section 17.2.

In stars with higher internal temperatures, particle energies are higher and there is a probability of a fusion process involving heavier nuclei where Coulomb barrier energies are greater. One process, known as the carbon–nitrogen–oxygen

Figure 13.4 | Contributions to the Proton–Proton Cycle in the Sun

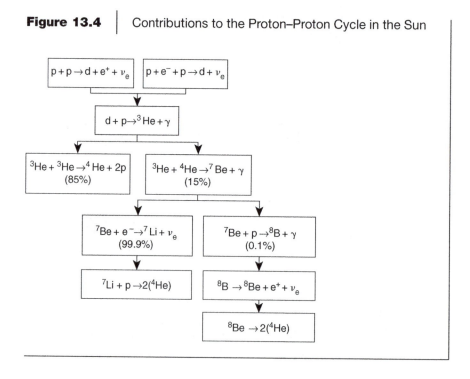

cycle (or CNO cycle), is equivalent to the proton–proton cycle since it ultimately fuses four hydrogen into one helium. It proceeds as follows:

$$^{12}C + p \rightarrow {}^{13}N + \gamma$$

$$^{13}N \rightarrow {}^{13}C + e^+ + \nu_e$$

$$^{13}C + p \rightarrow {}^{14}N + \gamma \qquad . \qquad (13.29)$$

$$^{14}N + p \rightarrow {}^{15}O + \gamma$$

$$^{15}O \rightarrow {}^{15}N + e^+ + \nu_e$$

$$^{15}N + p \rightarrow {}^{12}C + {}^4He$$

Again two β-decay processes are required to convert two of the protons to neutrons. Since carbon is neither created nor destroyed but merely acts as a catalyst, the Q for this process is the same as for the proton–proton cycle. Although this process requires greater energy to overcome the Coulomb barrier, it is not limited by the rate of deuteron production as is the proton–proton cycle. Hence, the CNO cycle is less likely at lower temperatures than the proton–proton cycle but becomes more probable at higher temperatures. This trend is indicated in Figure 13.5. The internal temperature of the sun, about 10^7 K, falls below the cross-over point indicating that the energy production in the sun is dominated by the proton–proton cycle.

Figure 13.5 | Relative Importance of the Proton–Proton and CNO Cycles in Stars as a Function of Their Internal Temperature

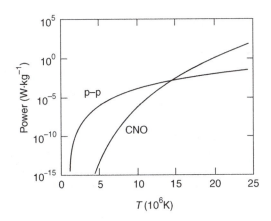

Curves are calculated from expressions for the temperature dependent reaction rates given in M. Schwarzschild, *Structure and Evolution of the Stars*. Princeton: Princeton University Press, 1958.

Stars that have depleted their supply of hydrogen can no longer produce energy by the proton–proton or CNO cycles. This frequently happens in the central region of a star where temperatures and hence hydrogen fusion rates are greater resulting in a core made up primarily of ^{4}He. If the temperature in this region is sufficiently high (greater than about 10^8 K) further energy can be produced by helium fusion:

$$^4\text{He} + {}^4\text{He} \rightarrow {}^8\text{Be} + \gamma. \tag{13.30}$$

The ^{8}Be is unstable and decays by α-decay:

$$^8\text{Be} \rightarrow {}^4\text{He} + {}^4\text{He}. \tag{13.31}$$

The lifetime for this process is 7×10^{-17} s. Although some (or even most) of the ^{8}Be that is formed decays back to ^{4}He, some will fuse with another ^{4}He to produce ^{12}C:

$$^8\text{Be} + {}^4\text{He} \rightarrow {}^{12}\text{C} + \gamma. \tag{13.32}$$

This process is known as the triple alpha process since it combines three ^{4}He or α-particles into a ^{12}C nucleus. It has a value of $Q = 7.27$ MeV.

Heavier nuclei can be synthesized in the interior of stars by a variety of processes including neutron capture and α-particle capture. An important feature of all such processes is that fusion is no longer energetically favorable for nuclei heavier than around $A = 55$ (see Figure 4.1). As a result the relative abundance in the universe of elements heavier than Fe is much less than that of elements lighter than Fe. A good summary of nuclear astrophysics and in particular nucleosynthesis (that is, the methods by which heavier nuclei are synthesized in stellar interiors) can be found in K. S. Krane, *Introductory Nuclear Physics*, and references therein.

13.4 FUSION REACTORS

It is interesting to consider the energy release from various processes. Table 13.1 gives the typical energy produced per kilogram of fuel for different reactions. A chemical reaction corresponds to the burning of a fuel such as coal or oil. As a rough approximation for fission power it has been assumed that all available fission energy can be extracted from the ^{235}U component of natural uranium and that there is no contribution from fission of ^{238}U. For fusion power it is assumed that the energy is produced by the indicated reaction from the naturally occurring hydrogen isotopes in water. It is clear in all cases that nuclear reactions produce more energy than chemical reactions. This is merely a result of the difference between electronic binding energies (a few eV) and nuclear binding energies (a few MeV). The substantial amount of energy produced by p–p fusion is the result of the large mass of protons contained in one kilogram of water combined with the small atomic weight of ^{1}H. The smaller amount of energy available from d–d fusion is primarily the result of the much lower natural abundance of ^{2}H.

Table 13.1 | Comparison of Energy Produced by Chemical Processes, Fission of Natural Uranium, and Fusion of the Proton and Deuteron Components of Natural Water

Fuel	Reaction	Mass (g) of Reactive Component per kg of Fuel	Energy (J) per kg Fuel
chemical	chemical	—	6×10^6
natural uranium	^{235}U fission	7.2	6×10^{11}
natural water	p–p fusion	110	7×10^{12}
natural water	d–d fusion	0.016	3×10^9

On the basis of Table 13.1 it would certainly be desirable to be able to extract energy from water by p–p fusion according to equation (13.2). However, as this reaction is dominated by the weak interaction its cross section is very small. Thus, although it is an important factor in the energy production in the sun, it is unsuitable for a fusion reactor. The d–d fusion reactions given by equations (13.5) and (13.6) are possible candidates for a fusion reactor. These two reactions produce similar amounts of energy and as indicated in Table 13.1, natural water contains sufficient quantities of deuterium to make this a highly attractive energy source. The energy dependence of the reaction rate for d–d fusion is illustrated in Figure 13.1. By comparison, the figure shows that for moderate energies the reaction rate for d–t fusion is about two orders of magnitude larger. For this reason current fusion reactor research deals with d–t fusion.

In order to gain energy from a sustained fusion reaction, it is essential that the fusion energy output be greater than the sum of the energy input required to heat the fuel and the energy losses in the system. The energy gained per unit time per unit volume (power per unit volume) from fusion reactions in a plasma can be expressed as

$$P_f = RQ \tag{13.33}$$

where R is the reaction rate given by equation (13.21), Q is the energy per fusion. The fusion power per unit volume can be determined from the values of $\langle \sigma v \rangle$ illustrated in Figure 13.3 and is shown as a function of plasma temperature for d–d and d–t fusion in Figure 13.6. Even in the ideal situation, a plasma will lose energy as a result of bremsstrahlung. This loss results primarily from the behavior of the electrons, which, being much less massive than the ions, experience greater acceleration during electronion interactions. The power radiated per unit volume is

$$P_r = \frac{4\pi n n_e Z^2 e^6}{3(4\pi\varepsilon_0)^3 c^3 \hbar} \left(\frac{3k_B T}{m_e^3} \right)^{1/2} \tag{13.34}$$

Figure 13.6 | Power per Unit Volume Produced by a Plasma from d–d and d–t Fusion and Power Loss per Unit Volume from Bremsstrahlung (broken line) Plotted as a Function of Energy

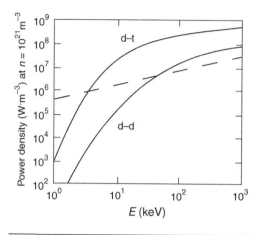

where n and n_e are the ion and electron densities and Z is the ion charge. The bremsstrahlung losses are the same for d–d and d–t plasmas (since Z is the same for both ions). The temperature dependence of these losses is compared with the fusion power in Figure 13.6. This makes clear the more demanding temperature requirements for d–d fusion compared with d–t fusion. In either case, it is necessary to operate the reactor at a temperature above the crossing point of these curves. It is now important to consider a comparison of fusion energy produced and the thermal energy needed to heat the plasma. From equations (13.21) and (13.33) the fusion energy per unit volume can be expressed as

$$E_f = \frac{n^2 \langle \sigma v \rangle}{4} Q \tau \tag{13.35}$$

where τ is the time during which the plasma is held at a temperature and density compatible with the fusion rate of equation (13.21). For simplicity we have considered d–d fusion where $n_1 = n_2 = n/2$. The thermal energy needed to heat the plasma is the sum of the energy needed to heat the electrons and the energy needed to heat the ions. In the case where $n_e = n$ then the total thermal energy needed per unit volume is

$$E_{th} = 3nk_B T. \tag{13.36}$$

For a reactor operating at a temperature above the bremsstrahlung crossing point, radiative losses can be ignored. Combining equations (13.35) and

(13.36) gives the minimum operating conditions to gain energy from the fusion process:

$$n\tau > \frac{12 k_B T}{\langle \sigma v \rangle Q}. \tag{13.37}$$

This condition is referred to as the Lawson criterion and the quantity on the left-hand side of the equation is generally called the Lawson parameter. The value of the Lawson parameter for which the reactor produces net energy depends on the operating temperature and the specific fusion reaction. However, equation (13.37) can be used as a general guideline for reactor design criteria and indicates that the achievement of a useful controlled fusion reaction requires that a sufficient plasma density be obtained for a sufficient period of time.

A major difficulty in constructing a fusion reactor is the requirement for a means of confining the plasma. Certainly the temperatures involved are sufficiently high that the interaction of the plasma with the walls of a containment vessel would readily melt any possible solid container material. It is also worth noting that any such interaction is also highly undesirable as it results in energy loss from the plasma. There are two confinement methods in common use in fusion reactor research—magnetic confinement and inertial confinement. Some details of these two methods are discussed below.

Magnetic Confinement   Since the ions and electrons in a plasma are free to move independently their motion can be controlled by the application of a suitable magnetic field. Magnetic confinement reactors utilize magnetic fields to direct the particles in a plasma to prevent them from colliding with the walls of the containment vessel. There are two basic geometries that are used for these devices—a linear geometry and a toroidal geometry.

The linear geometry uses a plasma column that is pinched at the ends. The plasma is contained in a cylindrical chamber and an axial magnetic field is provided by coils around the outside of the chamber. Basically the particles travel in a region of comparatively low field along the length of the cylinder and are reflected from the ends of the cylinder by regions of greater field. This is sometimes referred to as magnetic mirror confinement and the names for specific reactor types, Q-pinch, Z-pinch, etc., generally refer to the geometry of the mirror fields. In general, progress towards the conditions necessary for a sustained fusion reaction in a mirror confinement device has fallen short of that achieved in other reactor designs. As a result most current fusion research is directed towards the toroidal reactors and inertial confinement reactors described below.

The plasma column may be closed in the form of a toroid, in which case the particles travel along toroidal field lines produced by currents in windings around the toroid as illustrated in Figure 13.7a. The currents in this direction are referred to as poloidal currents. In this geometry the windings are closer together on the inside of the torus than on the outside resulting in a stronger magnetic field near the inside. The consequence of this is that the particles will slowly spiral outward, towards the region of weaker field and eventually strike the outer wall of the torus. In order to compensate for this effect an additional

Figure 13.7 | Geometry of Currents and Magnetic Field Lines in a Toroidal Reactor

(a) Toroidal field produced by poloidal currents and (b) poloidal field produced by a toroidal current.

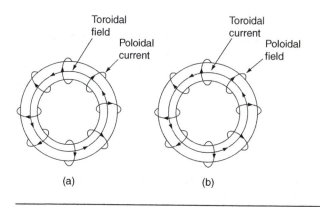

(poloidal) field is applied as illustrated in Figure 13.7b. The net field lines will be helical in shape and the particles will avoid interaction with the chamber walls as they follow the curved field lines. The poloidal field can be produced in one of two ways. Toroidal magnet windings can produce a toroidal current and hence a poloidal field. The device that operates on this principle is referred to as a *Stellerator*. An alternative approach is to induce a toroidal current in the plasma by treating the plasma as the secondary winding of a large transformer and this will produce the poloidal field. The device that operates on this principle is referred to as a *Tokamak* (an acronym for the Russian name of the device) and is shown in Figure 13.8. One should not necessarily think of a toroid with the

Figure 13.8 | Induction Method of Producing a Toroidal Field in a Tokamak

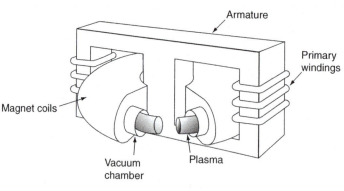

general shape illustrated in Figure 13.8 since many designs have a poloidal diameter that is not much smaller than the toroidal diameter. In fact, many designs do not have a poloidal cross section that is circular. One of the more successful designs has been the spherical Tokamak, which resembles a sphere with a circular hole through it (that is, the cored-apple geometry).

A major factor in the operation of a magnetic confinement reactor is the means by which the plasma is heated. The motion of the electrons and ions through the plasma generates heat in the same way a current flowing through a resistor produces thermal energy. This, however, is not sufficient to heat the plasma to the necessary temperature. Additional energy must be input into the reactor to raise the plasma temperature. The most common method of doing this is by neutral beam injection. A collection of atoms (typically deuterium) are ionized. The ions are accelerated to high energy by a small linear accelerator. These ions are then recombined with electrons to produce a high energy beam of neutral atoms. These neutral atoms are injected into the plasma where they are quickly ionized to produce ions and electrons. This process has two important features: to heat the plasma by transferring kinetic energy to other ions and electrons through coulombic interactions and to increase the density of the plasma. Both processes help to achieve the Lawson criterion. It is necessary to inject the particles as neutral particles because it would be difficult for charged particles to penetrate the magnetic fields surrounding the reactor.

Inertial Confinement   Inertial confinement refers to the situation where the fusion fuel is confined by inertial forces in the plasma itself. Most experiments that fall into this category are referred to as laser fusion experiments. A pellet of fuel (in most cases a mixture of deuterium and tritium contained in a capsule about a millimeter in diameter) is bombarded from several directions at once by high energy laser beams. The fuel pellet heats rapidly to a temperature that is, hopefully, suitable for fusion to take place. The actual processes that take place in the pellet as it heats are quite complex. Figure 13.9 shows a simplified description

Figure 13.9	Processes that Occur in a Fuel Pellet Bombarded by Laser Radiation

(a) Absorption of energy from laser beam, (b) formation of plasma atmosphere, (c) ablation of plasma atmosphere, and (d) compression of pellet core by ablation shock wave.

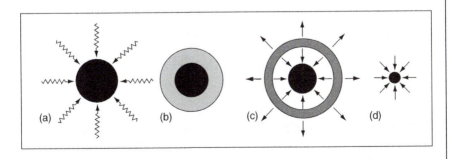

of this process. In Figure 13.9a, the laser radiation is absorbed by the fuel pellet, heating it from the outside. In Figure 13.9b the heat propagates through the pellet transforming the outer portions into a plasma. In Figure 13.9c this outer plasma atmosphere is driven off as it heats and expands. This process is referred to as ablation. In Figure 13.9d the remaining core of the pellet is compressed and heated by the inertial forces resulting from the expanding plasma atmosphere. The lasers used in such experiments are pulsed and the duration of the process shown in the figure is typically about 10^{-9} seconds. During this time the temperature of the fuel is very high because of the large amount of energy absorbed from the laser beam and the density of the pellet core can be compressed to densities of several thousand times the density of water. Thus, despite the small values of τ, the values of n can be sufficiently large that there is a possibility of achieving the Lawson criterion.

A major factor in the operation of an inertial confinement fusion reaction is the ability of the fuel pellet to absorb energy from the laser beam. Figure 13.10 shows the percent power absorbed by the d–t fuel pellet as a function of laser wavelength. High power lasers (for example, Nd-glass and CO_2) inevitably radiate at wavelengths around 10,500 Å. The figure shows that this is a disadvantageous situation. A crystal, known as a frequency doubler or second harmonic generator, can be used to reduce the wavelength of the radiation by a factor of two. This substantially increases the efficiency of energy absorption. From a quantum mechanical viewpoint one can think of the second harmonic generator crystal as combining two low energy photons to produce one high energy photon. From a classical standpoint, this property results from the nonlinear optical behavior of

Figure 13.10 | Percent of Laser Power Absorbed by a d–t Fuel Pellet as a Function of Laser Wavelength

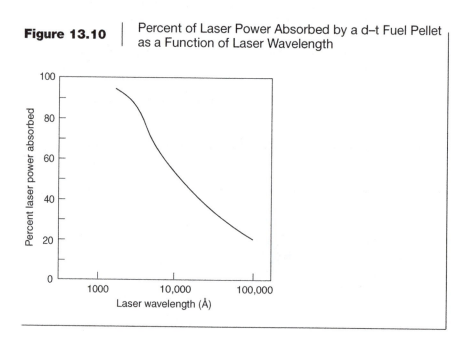

the crystal. The relationship between applied electric field and induced polarization is not linear, resulting in the production of higher harmonics in the radiated power spectrum. The greater the amplitude of the driving field the more intense the higher harmonics. Thus the second harmonic generation efficiency is negligible for weak incident radiation but can be substantial for intense laser light. The most commonly used second harmonic generation material is potassium dihydrogen phosphate (KDP) although new materials with greater efficiencies and greater power handling capabilities are now being developed.

13.5 PROGRESS IN CONTROLLED FUSION

In the present section the current status of some controlled fusion experiments will be summarized and the general philosophy for utilizing the energy produced by a fusion reactor will be discussed. An understanding of the meaning and implications of the Lawson parameter for the various reactor types is important. Plasma density and confinement time for inertial confinement fusion is a straightforward concept. It is clear that the plasma will achieve some maximum density for a well-defined period of time. The idea of confinement time for a Tokamak (for example) is, perhaps, not so obvious. However, the use of induction to generate the poloidal current requires the use of pulsed operating conditions. These pulses must, out of necessity, be quite long, as the plasma densities are substantially less than in the case of inertial confinement fusion. This kind of operation is sometimes referred to as quasi-steady state. A comparison of operating densities and confinement times for different types of d–t fusion reactors is shown in Figure 13.11. A convenient means of judging progress in fusion

Figure 13.11 | A Comparison of the Values of n and τ for Different Reactor Designs That Are Necessary to Achieve the Lawson Criterion for d–t Fusion

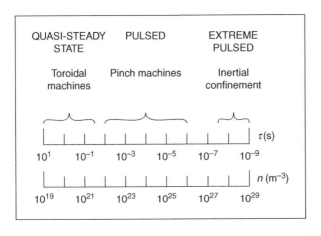

Figure 13.12 | Progress Toward an Operational Magnetic Confinement Fusion Reactor

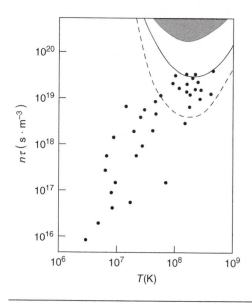

research is to plot the Lawson parameter as a function of plasma temperature. This is illustrated in Figure 13.12 for toroidal magnetic confinement reactors and in Figure 13.13 for inertial confinement reactors.

In Figure 13.12 the broken line represents unthermalized breakeven. This refers to the situation where the energy output of the reactor is equal to the energy input but the plasma conditions have been augmented by neutral beam injection. The solid line represents thermalized breakeven where the plasma conditions themselves are sufficient for net energy production. The shaded region represents ignition where the energy output is not only sufficient to yield a net energy gain but is also sufficient to maintain the plasma conditions. This is a self-sustained fusion reaction. These operating conditions refer to d–t fusion; conditions for d–d fusion would follow curves with values of $n\tau$ about two orders of magnitude larger. The data points in the figure represent the operating conditions of a number of experimental magnetic confinement reactors. The general trend of the points from the lower left to the upper right of the figure represents the chronological development of fusion reactors from the late 1960s to the late 1990s. This line also represents an increase in reactor power from the mW range to several MW. Present results are in the breakeven region and future developments can hope to achieve ignition. The time scale for such developments is presumably in the order of several decades.

In Figure 13.13 the solid region again represents ignition for a d–t reactor. The operating conditions for a number of inertial fusion reactors can be seen to

Figure 13.13 | Progress Toward an Operational Inertial Confinement Fusion Reactor

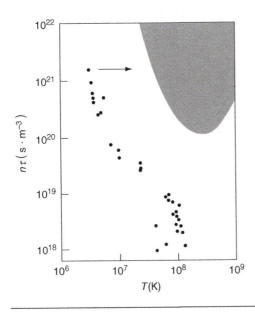

follow the trend from high-temperature, low-density reactors to low-temperature, high-density reactors. A new technique to ignite an already compressed fuel pellet with a second laser pulse can drive the operating conditions towards the ignition region as illustrated by the arrow in the figure. Experiments along these lines have shown promising results, although continued research efforts over many years will be required to produce an operational reactor.

Once a functioning experimental fusion reactor has been constructed it still remains to design and construct a viable reactor that can produce energy on a commercial scale. In this respect it is interesting to note that the first functioning experimental fission reactor was constructed in 1942 and the first commercial fission reactor became operational about twenty years later. An important consideration in the design of an operational reactor is the means by which the fusion energy may be converted into electricity. Utilizing the kinetic energy from fusion neutrons is not straightforward because of their low reaction cross section in most materials. Natural lithium consists of about 7% ^{6}Li and 93% ^{7}Li and is a useful material in this respect. Neutron reactions with lithium are

$$^6\text{Li} + \text{n} \rightarrow {}^3\text{H} + {}^4\text{He} \qquad (Q = 4.78 \text{ MeV}) \qquad (13.38)$$

$$^7\text{Li} + \text{n} \rightarrow {}^3\text{H} + {}^4\text{He} + \text{n} \qquad (Q = -2.47 \text{ MeV}). \qquad (13.39)$$

Figure 13.14 | Design of a Generic Fusion Reactor

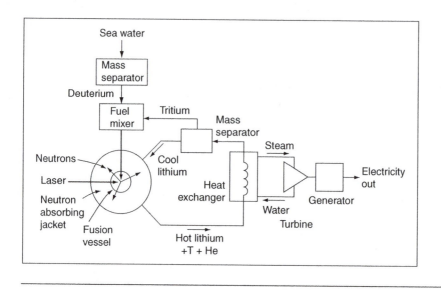

The first reaction is exothermic and has a large cross section. The second reaction has a smaller cross section and, although endothermic, is possible because the kinetic energy carried by the fusion neutrons is above the reaction threshold. Both reactions are breeder reactions as they produce tritium that can be recovered and utilized as reactor fuel. The only additional byproduct is ^{4}He. A generic d–t fusion reactor is illustrated in Figure 13.14. Kinetic energy of fusion neutrons is used to heat a lithium jacket around the reactor and is transferred to a steam turbine by a heat exchanger. It is important in such a reactor to properly deal with the fusion neutrons. Although these carry most of the fusion energy and can be used to breed new fuel, proper shielding is necessary to prevent the neutrons from being incident on materials where (n, γ) reactions could produce undesirable radioactive byproducts.

Although Table 13.1 illustrates the amount of energy that can be extracted from fusion fuel it does not fully address the question of the availability of fuel. Table 13.2 gives the number of years that world fuel reserves are expected to last. There are two substantial sources of uncertainty in the determination of these numbers. The first is the accuracy with which world energy requirements can be extrapolated. The second is in the estimation of the amount of fuel available. The uncertainty in this latter factor is quite large for fossil fuels but is small for d–d fusion. The amount of deuterium in the world's oceans is well known. In any case the precise values of the numbers in the table should not be taken too seriously. However, their implications concerning the potential importance of fusion energy are quite clear.

Table 13.2 | Estimated World Energy Reserves (in Years) from Different Power Sources

Energy Source	World Reserves (Years)
fossil fuels	200
fission (^{235}U nonbreeder reactor)	10^4
fission (^{235}U breeder reactor)	10^6
d–t fusion (^{6}Li breeder reactor)	10^7
d–d fusion (from sea water)	10^{10}

Problems

13.1. Justify the estimate of the energy produced by the combustion of 1 kg of fossil fuel as given in Table 13.1.

13.2. Estimate the Coulomb barrier height for two nuclei that are just in contact for the following pairs of nuclei: (a) d–d, (b) ^{6}Li–^{6}Li, and (c) ^{20}Ne–^{20}Ne.

13.3. (a) For d–d fusion in a plasma of temperature (energy) of 10 keV calculate the minimum Lawson parameter for sustained fusion.
(b) Repeat part (a) for d–t fusion.

13.4. (a) Calculate the Q for the process shown in equation (13.25).
(b) Calculate the Q for each of the processes shown in equation (13.29) and compare the total energy with the result of part (a).

13.5. For the ^{15}N(p, α)^{12}C reaction in equation (13.29), calculate the kinetic energies of the reaction byproducts.

13.6. The total energy output of the sun is 3.86×10^{26} W. Assuming that the energy comes exclusively from the proton–proton cycle, calculate the number of hydrogen nuclei involved in fusion reactions per second. Note that each neutrino has an average energy of 0.26 MeV.

13.7. (a) The reaction in equation (13.7) is the most likely candidate for a commercial fusion reactor. Calculate the mass of tritium needed to fuel a reactor that produces 5 GW of energy for a period of one year.
(b) If tritium is produced by the reactions (13.38) and (13.39) using natural lithium, calculate the mass of lithium required to produce the tritium mass calculated in part (a).

Particle Physics

Particles and Interactions

14.1 CLASSIFICATION OF PARTICLES

Further to the discussion in Chapter 2, we may divide all particles into two categories, fermions and bosons, as distinguished by their spins and the details of the statistics that describe their behavior. Figure 2.1 summarizes these two categories. Experimental evidence suggests that the leptons (electrons, neutrinos, etc.) are fundamental particles with no internal structure. In fact, results indicate that these are truly point particles. The next section of this chapter describes some of the properties of leptons in detail. The baryons and mesons show related internal structure and are collectively referred to as hadrons. The hadrons are not fundamental particles but are comprised of quarks. This will be discussed further in Chapters 15 and 16. A convenient way of categorizing particles for the purpose of the present discussion is illustrated in Figure 14.1. It is the last row of particles (leptons, quarks, and gauge bosons) that are fundamental. It is important to distinguish between particles that are fundamental and those that are stable. An electron is both fundamental and stable. A muon is fundamental but is not stable as discussed in the next section. The proton is stable (at least in the standard model—more on this in Chapter 17) but is not fundamental, while the neutron is neither fundamental nor stable.

Figure 14.1 | Classification of Particles

The last row represents those particles that are believed to be fundamental.

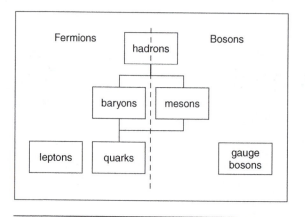

14.2 PROPERTIES OF LEPTONS

At present there are six known leptons and six corresponding antileptons as summarized in Table 14.1. It is seen that the leptons may be divided into three groups (called generations) each consisting of a lepton and an associated neutrino, as well as their antiparticles. There may be some uncertainty as to whether the particles given in Table 14.1 represent a comprehensive list of all possible leptons. Cosmological models of particle formation in the early universe indicate

Table 14.1 | Properties of Known Leptons

Lepton	Antiparticle	Generation	Mass (MeV/c^2)	Lifetime (s)	Decays
e^-	e^+	e	0.511	∞	—
ν_e	$\bar{\nu}_e$		~0	∞	—
μ^-	μ^+	μ	105.7	3.2×10^{-6}	$\mu^- \to e^- + \bar{\nu}_e + \nu_\mu$
ν_μ	$\bar{\nu}_\mu$		~0	∞	—
τ^-	τ^+	τ	1784	4.9×10^{-13}	$\tau^- \to e^- + \bar{\nu}_e + \nu_\tau$
					$\tau^- \to \mu^- + \bar{\nu}_\mu + \nu_\tau$
					$\tau^- \to \pi^- + \nu_\tau$
ν_τ	$\bar{\nu}_\tau$		~0	∞	—

that the number of lepton generations is limited to four or less. Some experimental evidence suggests that the three generations listed in the table, in fact, represent all leptons. Further discussions in the present text will assume this to be true.

The decay of the unstable leptons is an important topic. The best known of these is the decay of a muon (or its antiparticle) into an electron (or a positron) and two neutrinos. This can be written as

$$\mu^- \to e^- + \bar{\nu}_e + \nu_\mu. \tag{14.1}$$

We have seen in the previous discussion of β-decay processes that lepton number must be conserved in all processes. Equation (14.1) indicates that this conservation law is even more strict. We can define three separate lepton numbers corresponding to the three lepton generations, L_e, L_μ, and L_τ. The decay of the muon shows that (at least) L_e and L_μ must be conserved and it is this requirement that explains the existence of the two neutrinos on the right-hand side of the equation. It is the proper identification of neutrino generation that is the basis of our belief that lepton generation number (as well as total lepton number) must be conserved. As an example we can consider the β-decay of a proton:

$$p \to n + e^+ + \nu_e. \tag{14.2}$$

The neutrino can be moved to the left-hand side of the equation to represent the interaction of an antineutrino and a proton:

$$\bar{\nu}_e + p \to n + e^+. \tag{14.3}$$

The process in equation (14.3) is well known and has been extensively studied (although the cross section is very small). Experimental investigations of muon neutrinos sometimes involve the decay of pions. The basic properties of the pions (or pi-mesons) are given in Table 14.2. These particles will be discussed in further detail in the next chapter. The decay of a negative pion will produce antimuon neutrinos according to the process

$$\pi^- \to \mu^- + \bar{\nu}_\mu. \tag{14.4}$$

Table 14.2 | Properties of Experimentally Observed Pions

Particle	Charge (e)	Mass (MeV/c^2)	Spin	Lifetime (s)	Decay Products
π^+	+1	139.57	0	2.6×10^{-8}	$\mu^+ + \nu_\mu$
π^-	-1	139.57	0	2.6×10^{-8}	$\mu^- + \bar{\nu}_\mu$
π^0	0	134.96	0	8.7×10^{-17}	$\gamma + \gamma$

Experiments have utilized this process to look for reactions that violate conservation of lepton generation number, for example:

$$\bar{\nu}_\mu + p \to n + e^+. \tag{14.5}$$

In traditional experiments, no clear evidence for such reactions has been observed, although new observations will be discussed further in Section 17.3. The distinction between neutrinos and antineutrinos has also been demonstrated by the failure to observe reactions, such as

$$\nu_e + p \to n + e^+, \tag{14.6}$$

that would violate lepton number conservation. Further ideas concerning these matters will also be discussed in Chapter 17.

14.3 FEYNMAN DIAGRAMS

From a classical standpoint the interaction between particles (or any objects) is the result of a field. From a macroscopic standpoint the fields that we can experience directly are gravitational fields and electromagnetic fields. An important feature of a classical field is the fact that it can transfer energy and momentum from one object to another. The acceleration of a mass in a gravitational field is a well-known example of this phenomenon. From a quantum mechanical standpoint fields are quantized and the field quanta are bosons. As an example, light can be considered classically in terms of electric and magnetic fields. However, from a quantum mechanical standpoint a quantum of electromagnetic radiation is described as a photon. Along similar lines the electromagnetic interaction between two charges is described quantum mechanically as the exchange of a photon. Bosons that take part in interactions are referred to as gauge bosons and are said to mediate the interaction. In addition to the photon, that mediates the electromagnetic interaction, there are gauge bosons that mediate the other interactions in nature. The four known interactions and the gauge bosons that mediate them are given in Table 14.3.

Table 14.3 | Gauge Bosons Associated with the Four Known Interactions

The graviton is a hypothetical particle that mediates the gravitational interaction and has never been observed experimentally.

Interaction	Gauge Boson	Mass (GeV/c^2)	Spin	Acts on
strong	gluons	0	1	hadrons
electromagnetic	γ	0	1	charges
weak	W^+, W^-, Z^0	80.4, 80.4, 91.2	1	leptons and hadrons
gravity	graviton	0	2	masses

Table 14.4 | Symbol Convention for Different Particles in Feynman Diagrams

Particles	Symbol
leptons, baryons, real mesons	————————
photons	∿∿∿∿
W^+, W^-, Z^0, virtual mesons	– – – – – –
gluons	◯◯◯◯

Feynman diagrams are a convenient means of representing particle inter-actions. Although they can be used in a more quantitative sense to calculate cross sections, for example, this approach is beyond the scope of the present book. Here we will use these diagrams to indicate the relationship between particles in interactions and decay processes. Table 14.4 describes the symbols com-monly used to designate various types of particles in Feynman diagrams. As a simple example of a Feynman diagram, let's look at the coulombic interaction between two electrons as shown in Figure 14.2. The electrons are represented by straight lines with arrows indicating their direction of propagation. Time pro-gresses from left to right in the diagram. The gauge boson that mediates the inter-action (the photon) is represented by the wavy line. Since the photon is uncharged (and has zero lepton and baryon numbers) it can propagate in either direction. The figure shows that the two electrons interact via the coulombic interaction by exchanging a virtual photon. It should be noted that the arrows for antilep-tons (and other antifermions), as will be seen in some subsequent diagrams, are drawn in the opposite direction, as it is customary to view these particles as prop-agating backward in time.

Figure 14.2 | Feynman Diagram for an Electron–Electron Interaction Illustrating the Exchange of a Virtual Photon

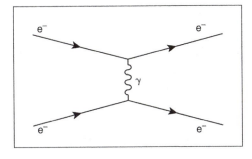

The conservation of mass/energy, linear momentum, angular momentum, lepton number, lepton generation number, baryon number, quark number, and charge applies to all particles in a reaction or decay. These conservation laws can be applied to the vertices of the Feynman diagram and the implementation of these laws is, for the most part, straightforward. However, the conservation of mass/energy and momentum requires some specific comments. In general, it is not possible to satisfy these conservation laws if the mass of the mediating particle is identical to the mass of the same particle when it is a free particle. Such particles do not obey the Einstein relation:

$$E^2 = p^2 m^2 + c^2 m^4, \qquad (14.7)$$

and are sometimes said to be "off the mass shell." Particles that satisfy equation (14.7) are referred to as real particles and those that do not are called virtual particles. In general, lines in Feynman diagrams that represent real particles have one free end and one vertex, while lines that represent virtual particles have vertices at both ends. The conservation of angular momentum also has some important implications for the construction of Feynman diagrams. The Feynman diagram shown in Figure 14.2 has vertices involving two fermion lines (leptons) and one boson line. Conservation of angular momentum requires that the vector sum of the spins of the particles must be a conserved quantity. An even number of half integer spins (fermions) is always necessary to compensate for an integer spin boson.

The construction of proper Feynman diagrams may be demonstrated by considering some of the decays and reactions that have previously been discussed in this book and an inspection of these diagrams is beneficial to understanding the physics of these processes. The Feynman diagram for the decay of a negative muon as described by equation (14.1) is shown in Figure 14.3. A W^- travelling downward (in the diagram) is equivalent to a W^+ travelling upward.

Figure 14.3 | Feynman Diagram for the Decay of a Negative Muon (Equation (14.1))

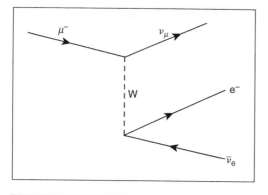

Conservation of lepton generation at vertices in the diagram requires that the weak interaction converts a muon to its own neutrino and creates an electron-antineutrino pair. It is, in fact, a general vertex rule that a charged weak gauge boson will convert a lepton into its own neutrino (or vice versa) or that it will create a lepton-antineutrino pair. The direction convention for antileptons can be seen in Figure 14.3. Vertex rules for gauge bosons will be discussed in a more general context in Chapter 16.

The Feynman diagrams for β-decay processes are illustrated in Figure 14.4. For β^-- and β^+-decay the weak gauge boson results in the creation of a lepton-neutrino pair. Equivalent diagrams can be drawn for W^+ and W^- with different directions of propagation. In each case charge conservation must be upheld at

Figure 14.4 | Feynman Diagrams for β-Decay Processes

(a) β^--decay (Equation (4.15)), (b) β^+-decay (Equation (4.17)), and (c) Electron Capture (Equation (9.3)).

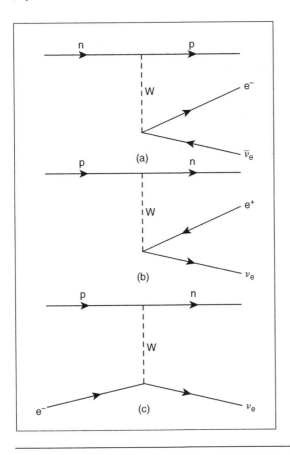

the vertices. A comparison of the diagrams for β^+-decay (equation (4.17)) and electron capture (equation (9.3)) illustrates an interesting feature of leptons. Moving a lepton from the right-hand side to the left-hand side of an equation converts it into its own antiparticle and results in a mirror reflection of the lepton line about the vertex in the Feynman diagram. An inspection of Figure 14.4 shows that the relevant conservation laws are still obeyed. The interaction of the weak charged boson with a baryon as shown in Figure 14.4 is best dealt with in the context of the quark components of the baryon and is left for Chapter 16. An attempt to construct valid Feynman diagrams for the processes given in equations (14.5) and (14.6) immediately demonstrates the relevant conservation law that is violated in each case.

Problems

14.1. The observation of the process given in equation (14.6) implies the existence of neutrinoless double β-decay. Explain how this is possible.

14.2. Following along the lines of Figure 14.4, construct diagrams for
  (a) Electron neutrino capture by a neutron
  (b) Positron capture by a neutron
  (c) Electron antineutrino capture by a proton

14.3. Calculate the Q values for the lepton decays shown in Table 14.1.

14.4. Unstable particles are often created in high energy collisions and the particles themselves are highly relativistic. It is interesting to consider the distance such a particle travels prior to decay. For the unstable leptons in Table 14.1 calculate the mean distance traveled by relativistic particles.

14.5. The lifetime of very short-lived states is sometimes given in energy units, that is the width in energy of the resonance corresponding to the state. Calculate the lifetime in time units of states with resonance widths of (a) 1 keV, (b) 1 MeV, and (c) 1 GeV.

14.6. Explain the observed trend for the pion lifetimes given in Table 14.2.

The Standard Model

15.1 EVIDENCE FOR QUARKS

The earliest versions of the quark model hypothesized the existence of three quarks (and three corresponding antiquarks). These were sufficient to explain the properties of the known hadrons at that time. Since both baryons and mesons are comprised of quarks it is obvious that quarks must be fermions. Conservation of angular momentum implies that it is possible to construct either fermions (baryons) or bosons (mesons) from a combination of fermions but it is not possible to construct fermions from a combination of bosons.

There is considerable experimental evidence that hadrons are made up of fundamental point-like particles that are, in many ways, analogous to the fundamental leptons. Although much of this evidence is rather indirect, taken as a whole it lends convincing support to the quark model. Here we review a small portion of the experimental evidence accumulated to date.

Neutral Meson Production Interaction of high energy electrons with protons is known to produce neutral mesons according to reactions such as

$$e^- + p \rightarrow e^- + p + \pi^0. \tag{15.1}$$

These reactions are difficult (at best) to explain if it assumed that the protons are, like electrons, fundamental structureless particles.

207

Figure 15.1 | Cross Section for Photons Incident on Protons

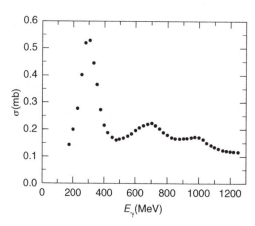

Data from T. A. Armstrong et al., *Phys. Rev. D5* (1972), 1640. Copyright 1972 by the American Physical Society.

Excited States of the Proton   The absorption cross section for photons on atoms shows anomalously large values at energies corresponding to those that allow for the population of excited states. The existence of excited states in atoms is a feature that results from the fact that an atom is a bound system of more than one particle. High energy photon absorption cross sections for protons show similar features indicating that the proton is not a discrete particle but is a bound system of more than one particle. Some experimental results are illustrated in Figure 15.1. The large peak at just below 300 MeV corresponds to the creation of the Δ^+ particle, a baryon with a mass of 1231 MeV/c^2 (that is, the proton mass, 938 MeV/c^2 plus the 293 MeV provided by the photon). Experimental results indicate that the Δ^+ is an excited state of the proton corresponding to a different spin state. Details of the distinction between the proton and the Δ^+ will be given in the next section.

Neutron Magnetic Moment   Although the neutron carries no charge, it does have a magnetic moment. This is a clear indication that the neutron has an internal structure involving a distribution of charges.

Deep Inelastic Scattering of Electrons   The most conclusive and revealing evidence for the existence of quarks comes from deep inelastic scattering experiments. These most commonly involve the scattering of high energy electrons by protons. These experiments follow from Rutherford's original α-particle scattering experiments to study the structure of the atom. Subsequent experiments using higher energy incident particles were used to study the structure of the nucleus.

The spatial resolution of a scattering experiment is given by the de Broglie wavelength of the incident particles:

$$\lambda = \frac{h}{p}. \tag{15.2}$$

Thus higher energy particles can probe the structure of the scatterer on a smaller scale. Electrons at an energy of 10 GeV have a de Broglie wavelength of about 0.1 fm (about one-tenth the radius of a proton). Scattering experiments utilizing electrons with energies up to 100 GeV can, therefore, probe the internal structure of a proton in considerable detail. The results of such experiments can be summarized as follows:

1. The proton contains three point-like particles.
2. These particles carry charges of $-1/3$ or $+2/3$ of the electron charge.
3. The particles are spin $1/2$ fermions.

These observations are consistent with the standard model of hadron structure as described in the remainder of this chapter.

15.2 COMPOSITION OF LIGHT HADRONS

The three quarks and three antiquarks of the original model are described in Table 15.1. From a quantum mechanical standpoint the identity of a quark (up, down, or strange) is referred to as the flavor of the quark. The baryons are comprised of three quarks while antibaryons are comprised of three antiquarks. The mesons are made up of a quark and an antiquark. It is, therefore, easy to see that the resulting baryons must be fermions and the mesons must be bosons. Even with only three quarks a substantial number of baryons and mesons can be constructed. This comes from the fact that quarks of different flavors can be

Table 15.1 | Properties of Quarks. T_3, S, C, B' and T' Represent Isospin, Strangeness, Charm, Bottomness, and Topness, Respectively (see Section 5.4)

The masses as shown are the current masses and are discussed further in the text.

Quark	Symbol	Charge (e)	Mass (GeV/c^2)	T_3	S	C	B'	T'
up	u	$+2/3$	0.0035	$+1/2$	0	0	0	0
down	d	$-1/3$	0.0061	$-1/2$	0	0	0	0
strange	s	$-1/3$	0.12	0	-1	0	0	0
charm	c	$+2/3$	1.35	0	0	$+1$	0	0
bottom	b	$-1/3$	5.3	0	0	0	-1	0
top	t	$+2/3$	176	0	0	0	0	$+1$

Table 15.2 | Properties of Light Mesons

Quarks	Spin 0 ($^1S_0, J^\pi = 0^-$)			Spin 1 ($^3S_1, J^\pi = 1^-$)		
	Particle	Charge (e)	Mass (MeV/c^2)	Particle	Charge (e)	Mass (MeV/c^2)
$\lvert u\bar{d}\rangle$	π^+	+1	140	ρ^+	+1	770
$\frac{1}{\sqrt{2}}\lvert d\bar{d} - u\bar{u}\rangle$	π^0	0	135	ρ^0	0	770
$\lvert \bar{u}d\rangle$	π^-	−1	140	ρ^-	−1	770
$\frac{1}{\sqrt{2}}\lvert d\bar{d} - u\bar{u}\rangle$	η	0	549	ω	0	783
$\lvert u\bar{s}\rangle$	K^+	+1	494	K^{*+}	+1	892
$\lvert d\bar{s}\rangle$	K^0	0	498	K^{*0}	0	892
$\lvert \bar{u}s\rangle$	K^-	−1	494	K^{*-}	−1	892
$\lvert \bar{d}s\rangle$	$\overline{K}^0$	0	498	$\overline{K}^{*0}$	0	892
$\lvert s\bar{s}\rangle$	η'	0	958	ϕ	0	1020

combined and that these bound systems can exist in various quantum mechanical states. Some examples help to illustrate these points.

Table 15.2 gives the properties of light mesons consisting of up, down, and strange quarks (along with their antiquarks). It is customary to indicate the quark content of the particle as a quantum mechanical state vector as given in the table. It is seen from the table that some particles (for example, π^0) are represented by a linear combination of quark flavor states.

The table illustrates some interesting relationships between the positively and negatively charged mesons and the neutral mesons. The process of *charge conjugation* interchanges quarks and antiquarks. Since the mass of a quark is the same as the mass of an antiquark of the same flavor, charge conjugation leaves the mass of a particle unchanged. The table indicates that charge conjugation changes positively charged mesons into corresponding negatively charged mesons of the same mass (or vice versa). These particles are the antiparticles of one another. The neutral mesons that are comprised of a quark and its own antiquark are said to be *self-conjugate* and are their own antiparticles.

Mesons have certain properties that are distinct from those of baryons that result directly from the fact that they are bosons. It is interesting to consider the properties of a bound quark–antiquark system in general. We can express the angular momentum of the quark–antiquark pair in terms of the three quantum numbers L, S, and J. In spectroscopic notation this is written as $^{2S+1}L_J$, where the usual convention S, P, D,... is used for $L = 0, 1, 2,...$. Since $\vec{J}$ is the vector sum of $\vec{L}$ and $\vec{S}$, the magnitude of J is constrained to take on values in the range

$$|L - S| \leq J \leq |L + S|. \tag{15.3}$$

Table 15.3 | Meson States for $L = 0$ and 1

Notation	L	S	J	π
$^1S_0$	0	0	0	−
$^3S_1$	0	1	1	−
$^1P_1$	1	0	1	+
$^3P_0$	1	1	0	+
$^3P_1$	1	1	1	+
$^3P_2$	1	1	2	+

Since the quark and the antiquark are spin $1/2$ fermions the total meson spin will be 0 for an antiparallel alignment of the quark spins or 1 for a parallel alignment of the quark spins. Some examples of angular momentum states for mesons are given in Table 15.3 where the values of L are 0 or 1, the values of S are 0 or 1, and the values of J are constrained by equation (15.3). Larger values of L are also possible, subject to the same constraints. These states are sometimes designated by their total angular momentum and their parity, π, as J^π. The question of the overall parity of a meson is quite interesting. The parity of a bound quark–antiquark pair with orbital angular momentum L is given by $(-1)^{L+1}$. The low-lying meson states are the least massive and those given in Table 15.2 all correspond to $L = 0$. As indicated in Table 15.3, the spin 0 mesons in Table 15.2 are in the $^1S_0$ (or 0^-) state, while the spin 1 mesons are in the $^3S_1$ (or 1^-) state. Mesons with the same quark content but different spins states are sometimes distinguished by appending a superscript asterisk to the name of the higher spin state particle to indicate an excited state; for example, K^{*0} vs K^0. Another scheme is to append the mass of the particle in MeV/c^2 to the name of the particle; for example, $\rho(1450)$ vs $\rho(770)$. The obvious increase in mass associated with increasing spin state as illustrated in Table 15.2 is evidence for a spin-spin interaction term in the potential.

The properties of baryons that can be formed from up, down, and strange quarks are given in Table 15.4. The three spin $1/2$ quarks can yield a total spin of $1/2$ (for two parallel and one antiparallel) or $3/2$ (for three parallel). Following the discussion above, the possible total angular momentum values of the three quark bound state are related to the spin and orbital components and some examples are illustrated in Table 15.5. The parity of an L state baryon is given by $(-1)^L$. The particles described in Table 15.4 are in the $L = 0$ state meaning that the spin $1/2$ baryons are in the $^2S_{1/2}$ (or $1/2^+$) state and the spin $3/2$ baryons are in the $^4S_{3/2}$ (or $3/2^+$) state. Finally, one should note that certain quark flavor combinations are not allowed in certain spin states, for example, the spin $1/2$ state of uuu.

Table 15.4 | Properties of Light Baryons

Quarks	Spin 1/2 ($^2S_{1/2}, J^\pi = 1/2^+$)			Spin 3/2 ($^4S_{3/2}, J^\pi = 3/2^+$)		
	Particle	Charge (e)	Mass (MeV/c^2)	Particle	Charge (e)	Mass (MeV/c^2)
$\lvert uuu \rangle$	—	—	—	Δ^{++}	+2	1230
$\lvert uud \rangle$	p	+1	938	Δ^+	+1	1231
$\lvert udd \rangle$	n	0	940	Δ^0	0	1232
$\lvert ddd \rangle$	—	—	—	Δ^-	−1	1234
$\frac{1}{\sqrt{2}}\lvert (ud - du)s \rangle$	Λ	0	1116	—	—	—
$\lvert uus \rangle$	Σ^+	+1	1189	Σ^{*+}	+1	1383
$\frac{1}{\sqrt{2}}\lvert (ud + du)s \rangle$	Σ^0	0	1192	Σ^{*0}	0	1384
$\lvert dds \rangle$	Σ^-	−1	1197	Σ^{*-}	−1	1387
$\lvert uss \rangle$	Ξ^0	0	1315	Ξ^{*0}	0	1532
$\lvert dss \rangle$	Ξ^-	−1	1321	Ξ^{*-}	−1	1535
$\lvert sss \rangle$	—	—	—	Ω^-	−1	1672

Antibaryons are formed from three antiquarks and can be viewed in terms of the application of charge conjugation to the baryons given in Table 15.4. In all cases the masses of the antibaryons are the same as the masses of the corresponding baryons while the charges are opposite (in the case of charged baryons). The parity of an antibaryon in an orbital angular momentum state L is given by $(-1)^{L+1}$. The neutral baryons are not self-conjugate and positive and negative

Table 15.5 | Baryon Angular Momentum States for $L = 0$ and 1

Notation	L	S	J	π
$^2S_{1/2}$	0	1/2	1/2	+
$^4S_{3/2}$	0	3/2	3/2	+
$^2P_{1/2}$	1	1/2	1/2	−
$^2P_{3/2}$	1	1/2	3/2	−
$^4P_{1/2}$	1	3/2	1/2	−
$^4P_{3/2}$	1	3/2	3/2	−
$^4P_{5/2}$	1	3/2	5/2	−

varieties of some baryons with similar (but not identical) masses exist that are not antiparticles of one another.

15.3 COMPOSITION OF HEAVY HADRONS

The description of the heavier mesons and baryons requires the introduction of additional, heavier quarks. Since the original version of the quark model, three additional quarks have been added: charm, bottom, and top, as indicated in Table 15.1. Some authors use the name beauty instead of bottom and truth instead of top.

A number of mesons can be formed that involve one or more heavy quarks (or antiquarks). Some of those that have been observed experimentally are summarized in Table 15.6. The masses given in the table are the spin 0 ground state masses. In most cases, more massive excited states are also observed. The $c\bar{c}$ state,

Table 15.6 | Properties of Mesons with Charm and Bottom

Masses are for ground state $^1S_0$ particles.

Quarks	Particle	Charge (e)	Mass (MeV/c^2)
$\lvert\bar{u}c\rangle$	D^0	0	1865
$\lvert u\bar{c}\rangle$	$\overline{D}^0$	0	1865
$\lvert\bar{d}c\rangle$	D^+	+1	1869
$\lvert d\bar{c}\rangle$	D^-	−1	1869
$\lvert\bar{s}c\rangle$	D_s^+	+1	1969
$\lvert s\bar{c}\rangle$	D_s^-	−1	1969
$\lvert c\bar{c}\rangle$	J/ψ	0	3097
$\lvert u\bar{b}\rangle$	B^+	+1	5279
$\lvert\bar{u}b\rangle$	B^-	−1	5279
$\lvert d\bar{b}\rangle$	B_d^0	0	5279
$\lvert\bar{d}b\rangle$	$\overline{B}_d^0$	0	5279
$\lvert s\bar{b}\rangle$	B_s^0	0	5369
$\lvert\bar{s}b\rangle$	$\overline{B}_s^0$	0	5369
$\lvert c\bar{b}\rangle$	B_c^+	+1	6400
$\lvert\bar{c}b\rangle$	B_c^-	−1	6400
$\lvert b\bar{b}\rangle$	Υ	0	9460

usually referred to as charmonium, was first observed in 1974. Almost simultaneously electron–positron collider experiments at the Stanford Linear Accelerator and experiments involving the collision of protons with light nuclei at Brookhaven National Laboratory yielded evidence for charmonium. The former researchers named the new meson, ψ, while the latter referred to it as J. Although the name controversy persisted for a number of years, current convention seems to be to call it J/ψ as indicated in the table.

Mesons containing the most massive of the quarks, top, were first reported from a series of experiments at Fermilab in 1994. However, theoretical prediction for the existence of the top quark predated these observations. The Fermilab experiments utilized collisions between high energy protons and antiprotons to produce a variety of massive hadrons. In a very small number of instances t$\bar{t}$ quark pairs are formed. The top quark decays to the bottom quark by the weak interaction on a time scale of 10^{-24} s. This decay produces mesons with bottom and lepton–neutrino pairs that can be observed experimentally. A careful analysis of the energy and momentum of decay byproducts allows for a reconstruction of the original top quark mass. (The meaning of quark masses will be discussed further in the next section). Figure 15.2 shows the reconstructed top quark mass from the small number of relevant events observed in the Fermilab experiments.

Figure 15.2 | Reconstructed Mass for the Top Quark

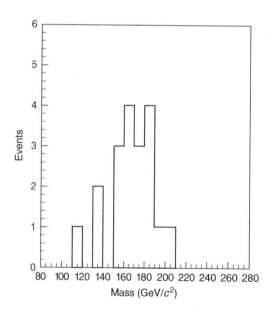

Data from F. Abe *et al.*, *Phys. Rev. Lett.* 74 (1995), 2626. Copyright 1995 by the American Physical Society.

The most recently discovered family of mesons contains both charm and bottom quarks. These were first reported in 1998 in experiments involving proton–antiproton collisions at Fermilab. These B_c mesons, which have a lifetime of 0.46 ps, were observed via the decay mode

$$B_c^+ \to J/\psi + \ell^+ + \nu_e \tag{15.4}$$

where ℓ is a positron or a muon. Similar decays are seen for B_c^-. Other possible decays for B_c include $J/\psi + \pi$, $B_s + \ell + \nu$, $B_s + \pi$ and a direct decay to leptons: $\tau + \nu_\tau$. In all cases these decays are the result of the weak interaction.

15.4 MORE ABOUT QUARKS

The standard model as described above assumes the existence of six quarks (and six corresponding antiquarks). These are divided into three generations and are in many ways analogous to the six known leptons. These relationships are summarized in Table 15.7. The leptons are either charge $-e$ or charge zero while the quarks are either charge $-e/3$ or charge $+2e/3$. Previous discussions have shown that lepton number as well as lepton generation must be conserved in all processes. Baryon number conservation can be viewed in terms of the quark content of the various hadrons. If we assign a baryon number of $+1/3$ to quarks and $-1/3$ to antiquarks then the quark content of baryons and mesons explains the assignment of baryon numbers to these particles. In a more fundamental sense baryon number conservation should be viewed as quark conservation. Quark flavor is clearly not conserved within a generation as β-decay processes represent the change of an up quark to a down quark, or vice versa. Changes in quark flavor are the result of the weak interaction and, as will be discussed further in the next chapter, changes in quark flavor between generation can occur.

Experimentally quarks have not been observed to exist as free particles and are said to be confined. *Confinement* results from the nature of the strong interaction. Although the strong interaction between nucleons in a nucleus is a short-ranged interaction, this is not a fundamental interaction between fundamental particles. The strong interaction between quarks (or between quarks

Table 15.7 | Relationship of the Generations of Leptons and Quarks

Antileptons and antiquarks are charge conjugates of the particles in the table.

	Leptons		Quarks	
Generation	Charge $-e$	Charge 0	Charge $-e/3$	Charge $+2e/3$
1	e^-	ν_e	d	u
2	μ^-	ν_μ	s	c
3	τ^-	ν_τ	b	t

and antiquarks), as mediated by gluons, actually increases with increasing distance. This is rather like stretching a spring. To get quarks further and further apart requires more and more energy; to actually separate them permanently would require infinite energy. It has been speculated that quarks and gluons may have existed in the very early universe (for the first 10^{-5} seconds or so) in the form of a quark–gluon plasma rather than in the form of hadrons. It is also possible that a quark–gluon plasma may exist at the center of a neutron star and experiments are underway to create this form of matter (for very short periods of time) in high energy particle collisions.

Because quarks are not observed as free particles, the concept of mass cannot be viewed in the same way as it is for particles such as electrons or protons. In the simplest approach the quark mass may be determined in terms of the mass of the hadrons that it comprises. The mass in this context is called the *effective mass* or *constituent mass*. This type of analysis would lead us to the conclusion that the masses of the up quark and the down quark were both about 310 MeV/c^2 with the down quark being more massive than the up quark by a small number of MeV/c^2. This simplistic approach is, more or less, consistent with the masses of many other hadrons, although it does not provide a measure of the mass of a quark if it existed as a free particle. Another approach is to view the quark masses as parameters in the theory of quark interactions (see Section 15.5 for more on this topic). Masses determined in such a way are referred to as *bare masses* (as they are, more or less, related to the intrinsic mass of the quark) or *current masses* (as the mass parameter appears in terms for currents in the theory). In the case of the light quarks, at least, the current masses are substantially less than the constituent masses. Some estimates of current masses for quarks are given in Table 15.1.

The relationships between the properties of the various hadrons can be understood in terms of quantum numbers associated with the flavors of their constituent quarks. Table 15.1 shows that quantum numbers for strangeness, charm, bottomness, and topness are defined for the respective quarks. Quantum numbers for the antiquarks are conjugates of those listed in the table. The up and down quarks are dealt with differently than other quarks. Table 15.1 indicates that these quarks are much less massive than the other quarks and as a result there are groups (or *multiplets*) of hadrons with very nearly the same masses that can be formed by interchanging up and down quarks. Tables 15.2 and 15.4 illustrate some examples of these multiplets; for example (n, p), (Σ^-, Σ^0, Σ^+), (π^-, π^0, π^+), etc. One view is to consider the up and down quarks as indistinguishable except for their difference in charge. As an analogy we can consider an atomic electron. The electron has an intrinsic spin of $1/2$. In the presence of a magnetic field electrons are designated as spin up or spin down on the basis of the z-component of their spin: $+1/2$ or $-1/2$, respectively. The flavor quantum number of the up and down quarks is defined as a pseudo-spin referred to as *isospin* with an intrinsic value of $T = 1/2$. In the presence of electromagnetic interactions the up and down quarks are distinguished on the basis of the *third component* of isospin, T_3, where the up quark is defined to have $T_3 = +1/2$ and the down quark is defined to have $T_3 = -1/2$. These can be related to the usual definitions of isospin of $+1/2$ and $-1/2$ for the proton and neutron, respectively.

Figure 15.3 | Eight-fold Way Diagrams for Light Mesons with (a) Spin 0 and (b) Spin 1

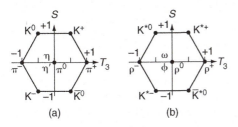

(a)   (b)

The relationships between different hadrons can be seen schematically in what are referred to as eight-fold way diagrams by plotting strangeness as a function of isospin. Examples for light mesons with spin 0 and spin 1 are illustrated in Figure 15.3. Nearly mass degenerate multiplets are indicated by families of particles on lines with constant strangeness. Eight-fold way diagrams for light baryons are illustrated in Figure 15.4. Similar diagrams that include heavier hadrons can also be constructed. For example, mesons and baryons with charm can be illustrated in a three-dimensional diagram with the charm of the particle plotted along the third axis. Because individual quarks have well-defined values for their various quantum numbers, as well as charge, the hadrons must also have certain relationships between these numbers. These relationships are defined by the Gell-Mann-Nishijima formula:

$$Q = T_3 + \frac{B + \Sigma(flavor)}{2} \tag{15.5}$$

where Q is the particle charge in units of e, B is the baryon number and $\Sigma(flavor)$ is the sum of flavor quantum numbers as defined in Table 15.1. Since individual quarks (and antiquarks) satisfy this relationship, all mesons and baryons must also obey the Gell-Mann-Nishijima formula. This is verified for all but the heaviest known hadrons by direct experimental evidence.

Figure 15.4 | Eight-fold Way Diagrams for Light Baryons with (a) Spin 1/2 and (b) Spin 3/2

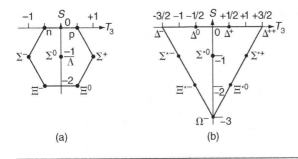

(a)   (b)

15.5 COLOR AND GLUONS

Since quarks, like electrons, are fermions they obey Fermi-Dirac statistics and are subject to the Pauli exclusion principle. A careful consideration of the properties of some baryons indicates a problem with the quark model as described above. For example, the Δ^{++} baryon is composed of three up quarks in the same spin and orbital states and this violates the Pauli exclusion principle. (This difficulty is not encountered for mesons because a quark and an antiquark are always distinguishable particles.) To resolve the problem of the Δ^{++} baryon, each of the three identical quarks in the system must be distinguished in some way. This is done by assigning an additional quantum number to each quark in such a way that they are distinguishable. This new quantum number is referred to as *color*, although this is in no way related to the conventional concept of color. The quantum number for the color of a quark can take on one of three values referred to as *red* (R), *green* (G), or *blue* (B). Antiquarks are assigned anticolor, which can have values of *anti*red ($\overline{R}$), *anti*green, ($\overline{G}$) or *anti*blue ($\overline{B}$).

Baryons are made up of three quarks of three different colors, that is, one red, one green, and one blue. Mesons are made of a quark of a given color and an antiquark of the same anticolor. This scheme has the net result that all hadrons are colorless. This is a requirement of the model since no manifestation of color is observed in the measured physical properties of mesons or baryons. This is in contrast to the flavor quantum numbers of the quark constituents of hadrons. A hadron can exhibit overall strangeness, charm and other characteristics resulting from the sum of the quantum numbers of the component quarks associated with these quantities. The physical manifestation of strangeness or charm is the mass of the hadron, since quarks of different flavors have different masses. The theory of interactions involving color is referred to as *Quantum Chromodynamics* (QCD) and will be dealt with only briefly in this text. However, a very phenomonological interpretation of color interactions can be made by analogy with electrostatic interactions. In the same way that like charges repel and unlike charges attract we can view the strong force between quarks as color dependent with like colors repelling and unlike colors attracting one another. Thus a meson, which contains a quark of a certain color, must contain an antiquark of the same anticolor in order to leave the meson as a color-neutral particle.

Quantum mechanically the properties of hadrons can be considered in terms of their wave functions and this approach illustrates the necessity for color. The total wave function of a particle is comprised of four components (space, spin, flavor, and color):

$$\psi = \psi_{space}\, \psi_{spin}\, \psi_{flavor}\, \psi_{color}. \tag{15.6}$$

For baryons (which are fermions) the total wave function is required to be antisymmetric under interchange of any two quarks. In this case it can be shown that the space-spin-flavor wave function is symmetric. The wave function

Figure 15.5 | Proton–Proton Scattering Viewed in Terms of (a) Virtual Pion Exchange Between Baryons and (b) Gluon Exchange Between Quarks

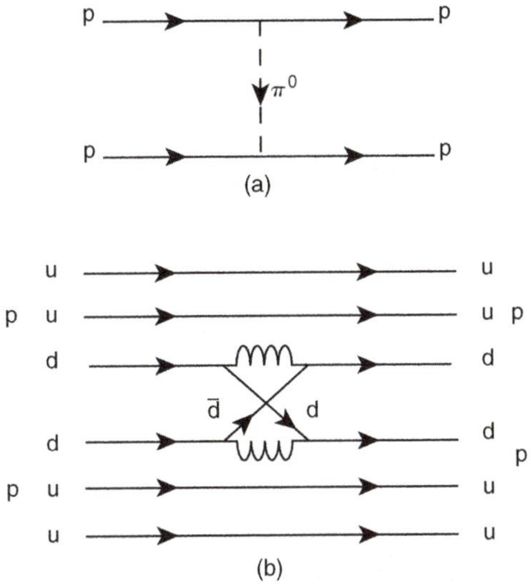

for the color state is defined by a linear combination of all possible color states:

$$\psi_{color} = \frac{1}{\sqrt{6}}[RGB + GBR + BRG - RBG - BGR - GRB]. \qquad (15.7)$$

It can be seen that this function is antisymmetric under interchange of any two quarks. When combined with the baryon space-spin-flavor wave function this restores the antisymmetric properties of the total wave function.

It is now possible to consider the details of the strong interaction between quarks. Virtual pion exchange as illustrated in Figure 15.5a is sometimes used to describe the interactions between nucleons in a nucleus. However, neither the hadrons nor the virtual pions in Figure 15.5a are fundamental particles, so the description of the interaction is not fundamental. An appropriate description of the interaction between hadrons in terms of the massless gluons that mediate the strong interaction is illustrated in Figure 15.5b. As mediators of the strong interaction, gluons cannot change the flavor of a quark. Thus the only allowed quark–gluon (q–g) vertices in Feynman diagrams are of the form $qg \rightarrow q$, $q \rightarrow qg$, $q\bar{q} \rightarrow g$, or $g \rightarrow q\bar{q}$ and their charge conjugates where the quarks must be of the same flavor. Figure 15.5b shows that in the quark description of the process, the virtual pion is represented by its quark components and results from the creation and annihilation of a $d\bar{d}$ pair.

The gluons form an octet with color state vectors given by:

$$|R\overline{G}\rangle$$

$$|R\overline{B}\rangle$$

$$|G\overline{R}\rangle$$

$$|G\overline{B}\rangle$$

$$|B\overline{R}\rangle \qquad\qquad . \qquad\qquad (15.8)$$

$$|B\overline{G}\rangle$$

$$\frac{1}{\sqrt{2}}|R\overline{R} - G\overline{G}\rangle$$

$$\frac{1}{\sqrt{6}}|R\overline{R} + G\overline{G} - 2B\overline{B}\rangle$$

Examples of the simplest cases of color exchange between quarks or between quarks and antiquarks of the same flavor are illustrated in Figure 15.6. Color is conserved at all vertices but because the gluon carries color, the color of the quark is changed. Color exchange between quarks in the context of reactions between hadrons requires that all mesons and baryons involved remain colorless. An example of color exchange process involving colored gluons for the reaction $p + \pi^+ \rightarrow \pi^0 + \Delta^{++}$ is illustrated in Figure 15.7. This is, of course, only one possible valid way of coloring the quarks and gluons. Note that all initial and

Figure 15.6 | Examples of Simple Gluon Color Exchange Interaction (a) Between Quarks and (b) Between a Quark and an Antiquark

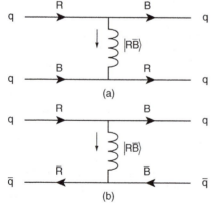

Figure 15.7 | Feynman Diagram for the Reaction $p + \pi^+ \to \pi^0 + \Delta^{++}$; (a) Showing Quark Relationships and (b) Showing Colored Gluons

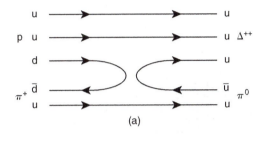

(a)

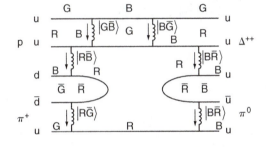

final hadron states are colorless. Note as well, that the intermediate Δ^{++} state is colorless and that the quark–antiquark creations and annihilations do not involve gluons or color exchange.

Problems

15.1. Show that the Gell-Mann-Nishijima relation is obeyed for the mesons given in Table 15.6.

15.2. Determine the isospin of the following particles: (a) π^-, (b) K^+, (c) D^0, (d) J, (e) Ξ^-, (f) n, and (g) Δ^{++}.

15.3. Draw a charm space eight-fold way diagram for spin 0 nonstrange mesons.

15.4. Draw Feynman diagrams showing quarks and gluons for the exchange of positive and negative pions between a neutron and a proton.

15.5. Draw Feynman diagrams for the τ^- lepton decays given in Table 14.1.

16 CHAPTER | Particle Reactions and Decays

16.1 REACTIONS AND DECAYS IN THE CONTEXT OF THE QUARK MODEL

With a knowledge of the quark content of baryons and mesons it is possible to view particle decays and reactions in terms of the fundamental quarks and leptons and the relevant interactions. As an example we can consider the β^--decay process shown in Figure 14.4 in terms of the quark components of the neutron and proton. This is illustrated in Figure 16.1 and this makes obvious the manner in which the weak interaction acts on quarks and on leptons. As we have seen previously the weak interaction changes a lepton to a neutrino or vice versa, but (in the context of the standard model) only within the same generation. Lepton–antineutrino creation or annihilation is an analogous process. Figure 16.1 shows that the weak interaction (unlike the strong interaction) can change the flavor of a quark. In fact the W^+ and W^- bosons must change the flavor of a quark. (We will discuss the properties of the Z^0 boson in more detail in the next section.) This is obvious, at least, from charge conservation, as the charged W boson must change a $+2e/3$ quark to a $-e/3$ quark (or vice versa), and similarly for the antiquarks. In the case of β-decay processes, the weak interaction changes a down quark to an up quark (as in Figure 16.1) or vice versa (for β^+-decay.)

Another example of a decay process is illustrated in Figure 16.2. This corresponds to the decay

$$D^0 \to K^- + \pi^+. \tag{16.1}$$

Figure 16.1 | Feynman Diagram for β^--Decay Showing Quark Relationships

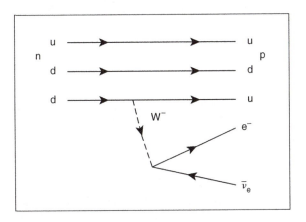

Here the weak $W^\pm$ boson couples two quarks and two flavor changes are necessary. In this case the $u\bar{u}$ quark pair is formed by the strong interaction. An alternate decay of the D^0 meson, illustrated in Figure 16.3, is given by

$$D^0 \rightarrow K^- + e^+ + \nu_e \tag{16.2}$$

and shows the coupling of the weak boson to leptons as in the case of β-decay. Figure 16.4 illustrates the decay

$$D^+ \rightarrow \overline{K}^0 + \pi^+ \tag{16.3}$$

Figure 16.2 | Feynman Diagram of the Weak Decay of a D^0 Meson to Hadrons (Equation (16.1))

In this diagram, and in all subsequent diagrams shown in this chapter that involve the strong interaction, the gluons are not shown.

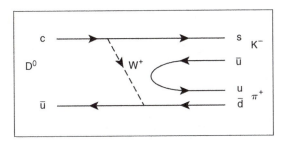

Figure 16.3 | Feynman Diagram of the Weak Decay of a D^0 Meson to a Hadron and Leptons (Equation (16.2))

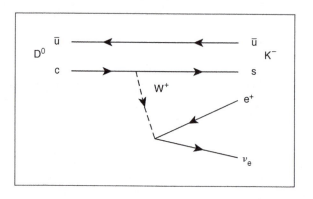

and shows that the weak W^+ boson must form quark–antiquark pairs of different flavors. In the examples given above there is no mixing of quark generation. Figure 16.5 shows an example of quark generation mixing in the K^--meson decay

$$K^- \to \mu^- + \bar{v}_\mu. \tag{16.4}$$

The implications of changes in quark generation will be discussed further in Section 16.3.

The above examples illustrate decays in which the weak interaction plays a role. Most interactions involving hadrons proceed dominantly by means of the strong interaction. The reaction

$$p + \pi^+ \to \Delta^{++} + \pi^0, \tag{16.5}$$

Figure 16.4 | Feynman Diagram of the Weak Decay of a D^+ Meson (Equation (16.3))

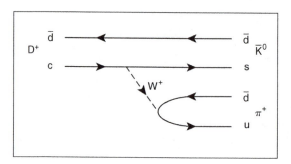

Figure 16.5 Feynman Diagram of the Weak Decay of a K⁻ Meson to Leptons (Equation (16.4))

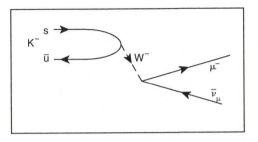

as illustrated in Figure 15.7 falls into this category. Here the reaction requires only the creation and annihilation of quark–antiquark pairs of the same flavor. In this sense both quark number as well as quark flavor is conserved as required for the strong interaction. A similar situation occurs in reactions such as

$$p + \pi^- \to n + \pi^0. \tag{16.6}$$

This is illustrated in Figure 16.6 and shows that there is merely an exchange of quarks between hadrons.

An example of a process that proceeds via the electromagnetic interaction is illustrated in Figure 16.7:

$$\Delta^+ \to p + \gamma. \tag{16.7}$$

Here a Δ^+ baryon decays to a proton by emitting a real photon. The quark content of the two baryons is the same but the masses are different. The emission

Figure 16.6 Feynman Diagram of an Example of the Strong Interaction Between Hadrons Showing Quark Exchange (see Equation (16.6))

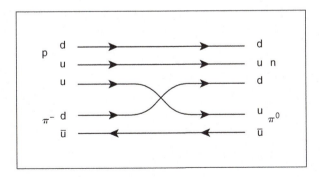

Figure 16.7 | Feynman Diagram of the Electromagnetic Decay of a Δ^+ Baryon (Equation (16.7))

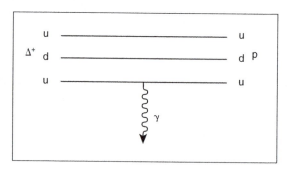

of the photon conserves mass/energy and corresponds to the flipping of one of the quark spins. Electromagnetic interactions can also be responsible for the creation or annihilation of particle–antiparticle pairs as shown in Figure 16.8. The particles may be leptons, as in e^-e^+ annihilation (Figure 16.8a), or quarks, as in the electromagnetic decay of a neutral pion (Figure 16.8b). In either case two (or more) real photons are produced as is necessary in order to conserve momentum.

Figure 16.8 | Feynman Diagrams for (a) Electron–Positron Annihilation to Photons and (b) Electromagnetic Decay of a ρ^0 Meson

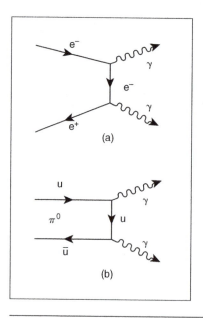

16.2 W± AND Z⁰ BOSONS

A variety of weak processes involving charged weak bosons have been discussed in previous sections. There is, as well, a neutral weak boson, the Z^0. Since the Z^0 boson does not carry charge it cannot mediate the kinds of processes that we have seen the $W^\pm$ bosons involved in. An example is the scattering of neutrinos by electrons as shown in Figure 16.9a and an example of neutrino–hadron scattering is shown in Figure 16.9b. Scattering involving only charged leptons and/or quarks (for example, electron–electron or electron–proton scattering) may be mediated by the Z^0 boson, although at low energies the electromagnetic interaction would be dominant. The Z^0 can also mediate particle–antiparticle creation or annihilation. Some examples are shown in Figure 16.10. Electron–positron annihilation as shown in Figure 16.10a:

$$e^- + e^+ \rightarrow \nu_e + \overline{\nu}_e, \tag{16.8}$$

requires the weak neutral boson, since the neutrinos cannot couple electromagnetically. However, at low energies the cross section for annihilation to real photons as shown in Figure 16.8a is greater. Electron–positron annihilation to hadrons, for example,

$$e^- + e^+ \rightarrow \rho^0 \tag{16.9}$$

(see Figures 16.10b and c) are generally mediated at low energies by the photon (Figure 16.10b) and at higher energies by the neutral weak boson (Figure 16.10c).

Figure 16.9 | Feynman Diagrams Showing Neutral Weak Bosons in (a) Neutrino–Electron Scattering and (b) Neutrino–Proton Scattering

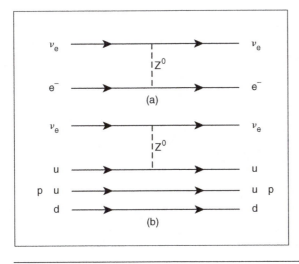

Figure 16.10 | Feynman Diagrams for Electron–Positron Annihilation to Neutrinos (a), to ρ^0 as Mediated by the Photon (b) and to ρ^0 as mediated by the Z^0 Boson (c). (Equations (16.8) and (16.9))

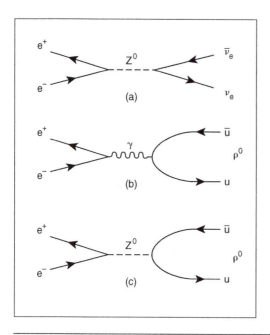

The above discussion shows that there are a number of similarities and some differences between the electromagnetic interaction and the weak interaction. The similarities are more obvious at high energies than at low energies. In fact, at sufficiently high energies the two interactions become completely equivalent. This feature is the basis of the *electroweak* theory developed by Glashow, Weinberg, and Salam that unifies the electromagnetic and weak interactions. At high energy the electroweak interaction is mediated by four massless gauge bosons consisting of a singlet (uncharged) and a triplet (with charges −e, 0, and +e). At lower energy the symmetry between the two interactions is broken and the singlet gauge boson remains massless and becomes the photon. The triplet particles acquire mass and become the weak $W^\pm$ and Z^0. This theory is able to predict the masses of the $W^\pm$ and Z^0 with a reasonable degree of accuracy and, in fact, made this prediction before the weak bosons were observed experimentally as free particles.

Thus far we have discussed the mediating weak bosons as virtual particles. The possibility of producing real particles in high energy collisions can be illustrated by the process of proton–proton scattering,

$$p + p \rightarrow p + p. \tag{16.10}$$

Figure 16.11 | Feynman Diagram for the Production of a Real π^0 Meson During Proton–Proton Scattering (Equation (16.11))

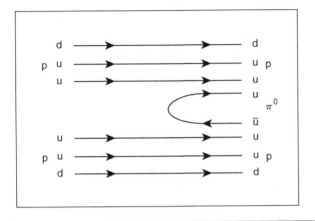

This can be viewed in terms of the quark components of the protons and the virtual pion as illustrated in Figure 15.5b. The production of a real neutral pion by the reaction

$$p + p \rightarrow p + p + \pi^0 \tag{16.11}$$

can occur at sufficiently high energy as illustrated in Figure 16.11. In a traditional accelerator experiment, a beam of protons may be incident on a target containing protons at rest. These target protons are often in the form of liquid hydrogen atoms that also serve as part of the bubble chamber used for particle detection. The kinetic energy of the incident beam required for real pion production substantially exceeds the rest mass energy of the pion that is produced. In the laboratory frame this is readily seen by the fact that not all of the incident kinetic energy is available for particle production, as conservation of momentum requires the scattered proton and the created pion to have kinetic energy. For a reaction $m_1 + m_2 \rightarrow m_3 + m_4 + m_5$, where a particle of mass m_1 is incident on a stationary (in the laboratory frame) particle of mass m_2, a relativistic derivation shows that the threshold kinetic energy is given by

$$K_{th} = \frac{(m_3 + m_4 + m_5)^2 c^2 - (m_1 + m_2)^2 c^2}{2 m_2}. \tag{16.12}$$

Utilizing the masses of the particles in equation (16.12), it is found that the production of a 135 MeV/c^2 neutral pion requires about 280 MeV of incident laboratory frame energy. Many contemporary particle experiments use a colliding beam geometry where conservation of momentum does not require that the particles on the right-hand side of the reaction have kinetic energy. As a result, the total kinetic energy of both beams becomes available for particle production and the advantages of this geometry become immediately obvious.

Figure 16.12 | Feynman Diagram for Real W⁺ Production During a Proton–Proton Collision (Equation (16.13))

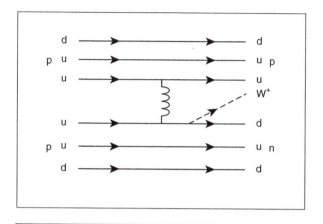

At sufficiently high energy proton–proton collisions can release a real weak boson by a process such as (see Figure 16.12):

$$p + p \rightarrow p + n + W^{+}. \tag{16.13}$$

Proton–antiproton collisions can produce real $W^{\pm}$ by quark–antiquark annihilation such as

$$p + \overline{p} \rightarrow \pi^{-} + \pi^{0} + W^{+} \tag{16.14}$$

and lepton collisions at sufficiently high energy can also lead to real $W^{\pm}$ production. An example is illustrated in Figure 16.13.

The first experimental observation of real weak boson production was in proton–antiproton collisions that produced W^{+} bosons. The charged weak bosons decay with a lifetime of about 3×10^{-25} s and are detected by the observation of their decay products. These can be lepton–neutrino pairs or quark–antiquark pairs as are observed for the decay of virtual $W^{\pm}$ in weak processes. These processes are summarized in Table 16.1. Conservation of lepton generation

Figure 16.13 | Feynman Diagram for Electron–Positron Annihilation to Real Weak Bosons

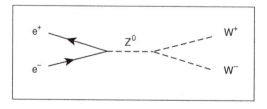

Table 16.1 | Decay Modes for the Real W⁺ Boson

Decay to	Branching Ratio (%)	Partial Width (GeV)
$e^+ v_e$	11	0.23
$\mu^+ v_\mu$	11	0.23
$\tau^+ v_\tau$	11	0.23
$u\bar{d}$	34	0.72
$c\bar{s}$	34	0.72

requires that the lepton and neutrino formed be from the same generation. This is not a strict requirement for quark–antiquark pair production. However, mixed quark generation decays have virtually zero branching ratios because of the very small cross sections for these processes (see more on this in Section 16.3). Decay to $t\bar{b}$ is not allowed for reasons of mass/energy conservation.

16.3 QUARK GENERATION MIXING

Experimentally it is known that the proton is the only stable baryon and there are no stable mesons. These facts can be readily explained by the existence of weak decays such as

$$\Sigma^- \to n + e^- + \bar{v}_e, \tag{16.15}$$

where there is a change of strangeness ($|\Delta S| = 1$). Similar generation mixing decays involving heavy flavors are also known. It is also possible for Σ^- to decay by the strangeness-conserving process

$$\Sigma^- \to \Lambda + e^- + \bar{v}_e. \tag{16.16}$$

where $|\Delta S| = 0$. On the basis of mass differences, and hence different Q values, we would expect that the branching ratios for these two processes would be different. However, these differences can be accounted for by following the discussion in Section 9.4 and we would expect that the two decays would have similar $f\tau_{1/2}$ values. Experimentally, however, it is found that

$$\frac{f\tau_{1/2}(|\Delta S| = 1)}{f\tau_{1/2}(|\Delta S| = 0)} \approx 12 \tag{16.17}$$

indicating that the strangeness-changing process can occur but is strongly inhibited. These observations can be explained in the context of the standard model by the following theory as originally proposed by Cabibbo.

The branching of decays is, in many ways, analogous to the flow of current into a node in a circuit as described by Kirchhoff's laws and the term current is often used in connection with this aspect of particle interactions. The total current, J, can be expressed as a linear combination of the currents representing the different decay modes. In the case of strangeness-conserving (J_0) and strangeness-changing (J_1) decays we can write

$$J = aJ_0 + bJ_1. \tag{16.18}$$

Since the transition rates are proportional to the square of the coefficients, normalization requires

$$a^2 + b^2 = 1. \tag{16.19}$$

It is customary to assume the form for the coefficients

$$
\begin{aligned}
a &= \cos\theta_C \\
b &= \sin\theta_C
\end{aligned} \tag{16.20}
$$

where θ_C is the Cabibbo angle and the condition given in equation (16.19) is satisfied. The Cabibbo angle can then be found to be

$$\theta_C \approx \cot^{-1}\sqrt{\frac{f\tau_{1/2}(|\Delta S| = 1)}{f\tau_{1/2}(|\Delta S| = 0)}}. \tag{16.21}$$

Following the example given above for Σ^--decay we obtain $\theta_C \approx 16°$.

Strangeness-changing decays exist because the weak interaction couples the up quark (u) to a linear combination of down (d) and strange (s) quarks, given by d′, rather than to just the down quark. Cabibbo theory gives

$$d' = d \cdot \cos\theta_C + s \cdot \sin\theta_C. \tag{16.22}$$

The charm quark (c) can be viewed as coupling to the linear combination

$$s' = -d \cdot \sin\theta_C + s \cdot \cos\theta_C. \tag{16.23}$$

This approach can be extended to include heavy flavors and, in general, explains the existence of generation mixing in weak processes.

16.4 CONSERVATION LAWS AND VERTEX RULES

On the basis of the information given in Chapters 14 and 15 and the previous two sections, it is possible to summarize the relevant conservation laws that apply at vertices in Feynman diagrams. In this section we summarize the types of vertices that are allowed for each type of interaction. We have seen that we can describe four kinds of gauge bosons: gluons (for strong interactions), photons (for electromagnetic interactions), $W^\pm$ (for charged current weak interactions), and Z^0 (for neutral current weak interactions). The weak bosons are divided into two categories for the purpose of the present discussion since different Feynman diagram vertex rules apply to the two types. The relevance of different conservation laws for the various gauge bosons acting on leptons and quarks are summarized in Table 16.2. As examples of allowed three-vertices in Feynman

Table 16.2 | Summary of Conservation Laws for Various Gauge Bosons Acting on Leptons and Quarks

Interaction	Gauge Bosons	Lepton $\leftrightarrow$ Neutrino	Quark Flavor Change	Quark Color Change
strong	gluons	no interaction	no change	yes
electromagnetic	photon	no change	no change	no change
weak	$W^{\pm}$	yes	yes	no change
weak	Z^0	no change	no change	no change

diagrams we can consider processes of the form

$$\text{fermion} \leftrightarrow \text{gauge boson} + \text{fermion} \qquad (16.24)$$

or those that correspond to creation/annihilation processes:

$$\text{fermion} + \text{fermion} \leftrightarrow \text{gauge boson}. \qquad (16.25)$$

Figure 16.14 illustrates examples of these vertices.

Figure 16.14 | Summary of Vertex Rules for the Interaction of Various Gauge Bosons with Leptons and Quarks

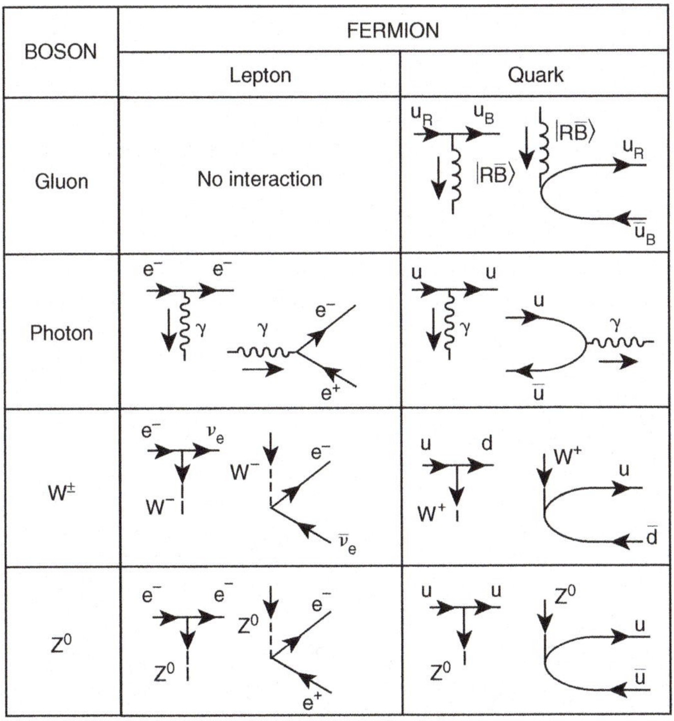

16.5 CLASSIFICATION OF INTERACTIONS

In the previous sections we have seen a wide variety of processes involving leptons and/or quarks and various gauge bosons. It is clear that processes involving leptons cannot be the result of the strong interaction as gluons do not couple to leptons (see Figure 16.14). However, a wide variety of other combinations of particle types and interactions are possible. We can categorize decay processes as described in Table 16.3. The relevance of the various interactions for each of these decays is best seen by examining the Feynman diagrams. These are illustrated in Figures 16.15 and 16.16. These examples should allow for the

Table 16.3 | Classification of Particle Decays

No.	Type of Decay	Particles	Gauge Bosons	Example
1	leptonic decay	lepton → leptons	weak	$\tau^+ \to e^+ + \bar{\nu}_\tau + \nu_e$
2	hadronic lepton decay	lepton → lepton + hadron	weak	$\tau^+ \to \pi^+ + \bar{\nu}_\tau$
3	hadronic decay	hadron → hadrons	strong	$\Delta^{++} \to \pi^+ + p$
4	leptonic hadron decay	hadron → leptons	weak	$\pi^+ \to \mu^+ + \nu_\mu$
5	semileptonic hadron decay	hadron → hadron + leptons	weak	$n \to p + e^- + \bar{\nu}_e$
6	nonleptonic hadron decay	hadron → hadrons	strong + weak	$K^+ \to \pi^+ + \pi^+ + \pi^-$
7	electromagnetic	hadron → hadron + photon	electromagnetic	$\Delta^+ \to p + \gamma$

Figure 16.15 | Examples of Lepton Decays as Given in Table 16.3

(a) and (b) correspond to numbers 1 and 2 in the table, respectively.

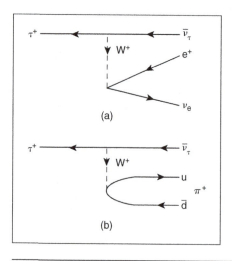

Figure 16.16 | Examples of Hadron Decays as Given in Table 16.3

(a) through (e) correspond to numbers 3 through 7 in the table, respectively.

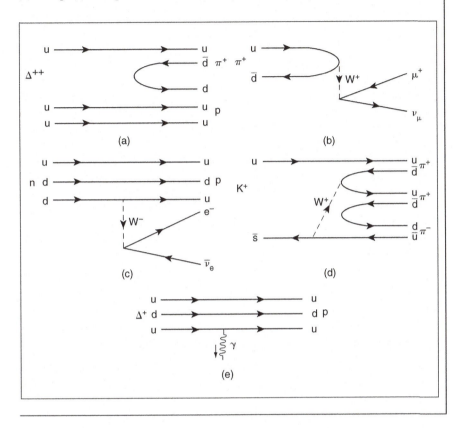

(a)

(b)

(c)

(d)

(e)

immediate classification of decays according to the type of initial and final particles as well as the type of gauge boson(s) involved.

16.6 TRANSITION PROBABILITIES AND FEYNMAN DIAGRAMS

Although a detailed analysis of reaction cross sections on the basis of Feynman diagrams is not possible here, we can make some approximate estimates of relative transition rates based on a qualitative analysis of the diagrams. In general, processes that are dominated by strong interactions proceed more rapidly. In situations such as Figure 16.16d, where both the strong interaction and the weak interaction are required, the lifetime is determined by the weak interaction as this determines the maximum rate at which the process can proceed. We have already mentioned that electron–positron annihilation via the electromagnetic interaction (Figure 16.8a) is more likely than via the weak interaction (Figure 16.13) at low energies. Along similar lines, we can compare neutral pion decay via the electromagnetic interaction (Figure 16.8b) and charged pion decay via the weak interaction (Figure 16.16b). The former decay has a lifetime of 8.7×10^{-17} s and the latter decay has a lifetime of 2.6×10^{-8} s.

A careful inspection of Feynman diagrams allows for a more detailed analysis of relative transition rates for a variety of hadronic processes that involve the strong and/or weak interactions. We begin with a consideration of weak processes. Figure 16.1 shows a well-known weak process, β-decay, which involves both leptons and quarks. Figures 16.2 and 16.4 show weak D-meson decays that do not involve leptons. In all of these cases there is no mixing of quark generation and the decays are not inhibited. As we saw in Section 16.3, processes involving quark generation changes are suppressed relative to generation-conserving processes. An example of a process that involves leptons and quark-generation mixing, the decay of the K^- meson, is shown in Figure 16.5. The decay of a K^+ meson, as shown in Figure 16.16d, does not involve leptons but has a quark generation change. In some cases, a weak boson that couples to two quarks can result in two generation changes and is highly suppressed. An example is the decay

$$A^+ \to \pi^+ + n, \tag{16.26}$$

as illustrated in Figure 16.17. From these trends we can establish an approximate hierarchy of transition probabilities for the weak decay of hadrons based on the number of quark generation changes that are involved.

Other suppressed situations occur, such as

$$\Sigma^+ \to n + e^+ + \nu_e \tag{16.27}$$

Figure 16.17 | The Decay of the A^+ Baryon Showing Generation Mixing for Two Quarks

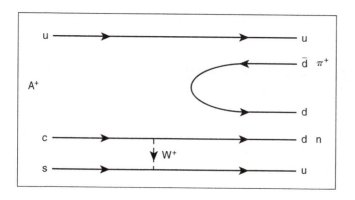

Figure 16.18 | The Decay of the Σ^+ Baryon Showing Two Weak Bosons

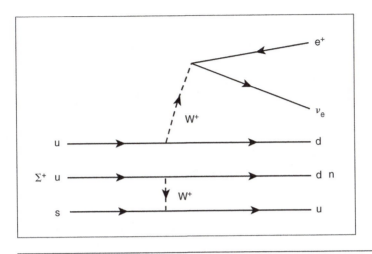

(see Figure 16.18), where two weak bosons are required in order to create the lepton–neutrino pair and to produce the necessary quark flavor changes.

A similar analysis can be considered for strong processes. As an example we consider the decay for the ϕ meson ($s\bar{s}$). The two principal decay modes for ϕ are decay to pions:

$$\phi \rightarrow \pi^+ + \pi^- + \pi^0 \tag{16.28}$$

and decay to K mesons (kaons):

$$\phi \rightarrow K^+ + K^-. \tag{16.29}$$

An inspection of meson masses indicates that the Q for the former process is 605 MeV while for the latter process it is 32 MeV. Our previous discussions suggest that the branching ratio for these decays would significantly favor decay to pions. However, experimentally it is found that the branching ratios are about 17% for decay to pions and 83% for decay to kaons. The Feynman diagrams as shown in Figures 16.19 and 16.20 are informative for understanding this behavior. Diagrams such as that shown for decay to pions are referred to as disconnected diagrams since there is no flow of quarks from the left to right sides of the diagram. *Zweig's rule* states that disconnected processes are suppressed relative to connected processes. To understand the reasons for this we must consider the transfer of color from the initial to final quarks, realizing that all meson states must be colorless. In Figure 16.20 quark color is transferred by

Figure 16.19 | The Decay of a ϕ Meson to Pions

Gluons are not shown but are discussed in the text.

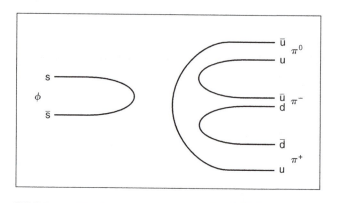

the $s\bar{s}$ quarks on the left side of the diagram to the strange quarks on the right side of the diagram. Color changes can be associated with individual gluon interactions along the lines of those shown in Figure 15.7. In Figure 16.19, however, individual quarks cannot carry color from the left side to the right side of the diagram. Since individual gluons do carry color, a colorless interaction requires two (or more) gluons. This is analogous to the suppression of weak processes that require two (or more) weak bosons.

Figure 16.20 | The Decay of a ϕ Meson to Kaons

Gluons are not shown but are discussed in the text.

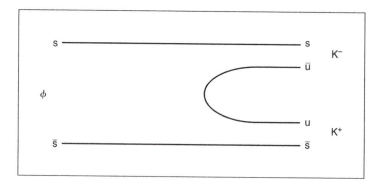

16.7 MESON PRODUCTION AND FRAGMENTATION

The study of meson production in high energy collisions provides important insight into the physics of particle interactions. Let's consider the production of particles in electron–positron collisions. One possible process is the production of lepton–antilepton pairs. At moderate energies this is most likely muon production:

$$e^- + e^+ \rightarrow \mu^- + \mu^+. \tag{16.30}$$

Another possible process is hadron production, that is, the production of quark–antiquark pairs (of the same flavor) as shown in Figure 16.10b. We would expect that when the total center of mass energy before the collision is equal to the rest mass energy of the quark–antiquark pair after the collision, then there would be a preference for hadron production relative to lepton production. A common experimental technique is to plot the ratio of cross sections for hadron production and muon production as shown in Figure 16.21. Resonances corresponding to the formation of several meson states are illustrated in the figure. The lowest energy states correspond to various $u\bar{u}/d\bar{d}$ states followed by the ϕ resonance corresponding to $s\bar{s}$ and, at higher energies, resonances involving heavy flavors. The ordering of states in this figure and their description in terms of the quark components of the corresponding mesons is generally considered to be confirmation of the quark model. Figure 16.21 also illustrates calculated cross sections with and without color. The large value of the cross section between the resonances, as shown in the figure, provides strong evidence for the existence of color.

In this type of experiment we would expect quark–antiquark states to be formed from quarks of the same flavor. However, in practice a wide variety of meson

Figure 16.21 | The Ratio, R = Cross Section for Hadrons/Cross Section for Muons, for Electron–Positron Collisions

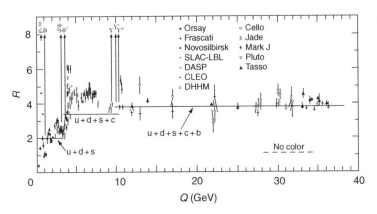

From F. Halzen and A. D. Martin, *Quarks and Leptons: An Introductory Course in Modern Particle Physics.* New York: Wiley, copyright 1984. Reprinted by permission of John Wiley & Sons, Inc.

Figure 16.22 | Feynman Diagram for the Process
$e^- + e^+ \rightarrow D^0 + \pi^0 + \pi^+ + D^-$

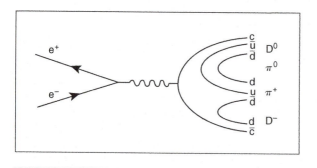

states of mixed flavor are typically observed. The reasons for this behavior have important implications for our understanding of the fundamental properties of gluons and quarks. Let's consider, for example, the formation of $c\bar{c}$ during an electron–positron collision. A possible observation might show the process:

$$e^- + e^+ \rightarrow D^0 + \pi^0 + \pi^+ + D^-. \tag{16.31}$$

We can understand how this can come about by looking at the Feynman diagram as shown in Figure 16.22. The gauge boson, shown as a photon, can also be a Z^0. Particle formation results from quark–antiquark pair formation. In this case there is no flavor mixing and the process on the right side of the gauge boson is the result of the strong interaction. We can view this process, called *fragmentation*, in terms of the breaking of gluons and the formation of new particle pairs as illustrated in Figure 16.23. Breaking a gluon bond produces new

Figure 16.23 | Representation of the Physical Processes Involved in the Breaking of Gluon Bonds in the Reaction Shown in Figure 16.22

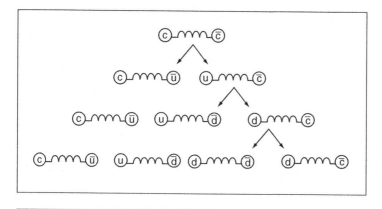

quarks and antiquarks at the ends of the broken bonds as shown in the figure. In this way a large variety of mixed flavor mesons can be formed.

16.8 CP VIOLATION IN NEUTRAL MESON DECAYS

We saw in Chapter 9 that β-decay can violate conservation of parity. This is because the parity operation (P) does not leave the system invariant. If the parity operation is combined with charge conjugation (C), then it can be shown that the β-decay process is in fact invariant. Thus, although β-decay can violate parity conservation it leaves CP unchanged.

The decay of certain neutral mesons demonstrates that the weak interaction can violate CP conservation. This has been observed in the decay of K^0 ($d\bar{s}$) and its antiparticle $\overline{K}^0$ ($\bar{d}s$) as discussed below and should apply as well to the decay of D^0, B^0_d, and B^0_s. The K^0 and $\overline{K}^0$ mesons can decay by several modes. These modes are to two pions, to three pions, and to a pion and leptons. The two pion modes are

$$K^0 \rightarrow \pi^0 + \pi^0 \tag{16.32}$$

and

$$K^0 \rightarrow \pi^+ + \pi^-, \tag{16.33}$$

and similarly for $\overline{K}^0$

$$\overline{K}^0 \rightarrow \pi^0 + \pi^0 \tag{16.34}$$

and

$$\overline{K}^0 \rightarrow \pi^+ + \pi^-. \tag{16.35}$$

Since both the K^0 and $\overline{K}^0$ mesons decay to the same daughter products, it is possible for processes such as

$$K^0 \rightarrow 2\pi \rightarrow \overline{K}^0 \tag{16.36}$$

to occur, which will change a K^0 to its antiparticle. In order to consider the invariance of CP for these processes let's look at the properties of the pions. Charge conjugation as applied to the wave functions of the pions has the following properties

$$C\psi(\pi^0) \rightarrow \psi(\pi^0)$$
$$C\psi(\pi^+) \rightarrow \psi(\pi^-) \tag{16.37}$$
$$C\psi(\pi^-) \rightarrow \psi(\pi^+)$$

and the parity operation is written as

$$P\psi(\pi^0) \rightarrow -\psi(\pi^0)$$

$$P\psi(\pi^+) \rightarrow -\psi(\pi^+) \tag{16.38}$$

$$P\psi(\pi^-) \rightarrow -\psi(\pi^-).$$

For a two-pion state, $\pi^0 + \pi^0$ or $\pi^+ + \pi^-$, with $L = 0$ it can be shown that the CP operation leaves the wave function invariant. That is

$$CP\psi(\pi^0, \pi^0) \rightarrow \psi(\pi^0, \pi^0)$$

$$CP\psi(\pi^+, \pi^-) \rightarrow \psi(\pi^+, \pi^-). \tag{16.39}$$

From equation (16.36) it follows that if the intermediate pion state is CP invariant then the initial and final meson states must also be CP invariant. However, applying the CP operation we find

$$CP\psi(K^0) \rightarrow \psi(\overline{K}^0)$$

$$CP\psi(\overline{K}^0) \rightarrow \psi(K^0) \tag{16.40}$$

and CP conservation is violated. Because, according to equation (16.36) K^0 can change into $\overline{K}^0$ and vice versa, the K^0 states are never pure but are admixtures of K^0 and $\overline{K}^0$. If we write these mixed states as

$$\psi(K_S) = \frac{1}{\sqrt{2}}[\psi(K^0) + \psi(\overline{K}^0)] \tag{16.41}$$

and

$$\psi(K_L) = \frac{1}{\sqrt{2}}[\psi(K^0) - \psi(\overline{K}^0)] \tag{16.42}$$

it can be shown that

$$CP\psi(K_S) = \psi(K_S) \tag{16.43}$$

and

$$CP\psi(K_L) = -\psi(K_L). \tag{16.44}$$

The meaning of the subscripts S and L will be discussed shortly. It is now clear that the mixture of states given by (16.41) for K_S is consistent with CP conservation for the two-pion decay mode. In order to understand the properties of the K_L state, let's look at the three-pion decay mode. This is

$$K^0 \rightarrow \pi^0 + \pi^0 + \pi^0 \tag{16.45}$$

or

$$K^0 \rightarrow \pi^0 + \pi^+ + \pi^- \tag{16.46}$$

and similarly for the $\overline{K}^0$. It can be shown that the CP operation as applied to the three-pion state gives

$$CP\psi(3\pi) = -\psi(3\pi). \tag{16.47}$$

Thus the three-pion decay mode for K_L is consistent with CP invariance. Experimentally K^0 mesons (which are a mixture of K^0 and $\overline{K}^0$ states) are found to have two well-defined lifetimes, 8.9×10^{-10} s and 5.2×10^{-8} s, corresponding to the decay of K_S and K_L. (The subscripts, therefore, refer to short- and long-lived states). It is also observed that the decay modes for these states are $K_S \rightarrow 2\pi$ and $K_L \rightarrow 3\pi$, consistent with CP invariance. (K_L can also decay to a pion plus leptons.) From a consideration of masses, the decay to three pions will have a smaller value of Q than the decay to two pions and it is, therefore, apparent that the three-pion decay will have a longer lifetime.

In 1964 an experiment conducted by Cronin and Fitch and coworkers looked for the decay $K_L \rightarrow 2\pi$, which would violate CP invariance. A careful analysis of their data indicated that, indeed, about 0.3% of the K_L decays were to two-pion states. These results provided clear evidence that weak processes can violate CP conservation.

Just as the inclusion of charge conjugation in the β-decay experiments described in Chapter 9 restored the conservation laws, there is common belief that inclusion of time reversal (T) for weak processes will restore the conservation laws; that is, these processes are invariant under CPT operations. Experimental evidence thus far does not contradict this hypothesis.

Problems

16.1. Some of the weak charmed decay modes of the bottom meson, B^+ are
    (a) $B^+ \rightarrow \overline{D}^0 + \mu^+ + \nu_\mu$
    (b) $B^+ \rightarrow \overline{D}^0 + \pi^+$
    (c) $B^+ \rightarrow D^- + \pi^+ + \pi^+$
    (d) $B^+ \rightarrow D_S^{\ +} + \overline{K}^0$
    Draw Feynman diagrams for these decays.

16.2. The decay

$$\overline{D}^0 \rightarrow K^- + \mu^+ + \nu_\mu$$

is inhibited while the decay

$$\overline{D}^0 \rightarrow K^+ + \mu^- + \overline{\nu}_\mu$$

is not. Explain.

16.3. In addition to the decay modes of the τ^- lepton given in Table 14.1, modes involving other combinations of light and strange mesons are possible. Discuss these and show Feynman diagrams as appropriate.

16.4. Discuss the possibility of decay modes of the W⁺ boson other than those shown in Table 16.1. Consider, in particular, decays to $u\bar{s}$, $u\bar{b}$, $c\bar{d}$, $c\bar{b}$, $t\bar{d}$, $t\bar{s}$, and $t\bar{b}$.

16.5. Draw a Feynman diagram for the process given in equation (16.14).

16.6. Draw Figure 16.2 indicating a valid coloring scheme for the quarks. Show gluons as necessary.

16.7. Discuss the possible decay modes of a real Z^0 boson.

16.8. Determine if each of the following decays is allowed. If it is allowed determine the interaction by which it proceeds. If it is not allowed determine which conservation law(s) is/are violated.

(a) $\pi^+ \rightarrow \pi^0 + e^+ + \nu_e$

(b) $\tau^+ \rightarrow \pi^+ + \bar{\nu}_\tau$

(c) $\Xi^- \rightarrow \Lambda + \pi^-$

(d) $\mu^- \rightarrow \pi^- + \nu_\mu$

(e) $\pi^+ \rightarrow e^+ + \nu_e$

(f) $\pi^0 \rightarrow e^+ + e^-$

(g) $\varphi \rightarrow K^+ + K^-$

(h) $\Omega^- \rightarrow \Xi^0 + K^-$

(i) $K^- \rightarrow e^- + \nu_e$

(j) $D^0 \rightarrow \Lambda^- + e^+ + \nu_e$

16.9. Determine if each of the following reactions is allowed. If it is allowed determine the interaction by which it proceeds. If it is not allowed determine which conservation law(s) is/are violated.

(a) $\Delta^+ + p \rightarrow p + p + \gamma$

(b) $\nu_e + \tau^- \rightarrow e^- + \nu_\tau$

(c) $K^+ + n \rightarrow p + \pi^0$

(d) $\Delta^{++} + \pi^+ \rightarrow K^+ + K^+$

(e) $p + \bar{p} \rightarrow \pi^+ + \pi^- + \pi^0$

(f) $\nu_\mu + \mu^- \rightarrow e^- + \nu_e$

(g) $\nu_\tau + e^- \rightarrow \nu_\tau + e^- + e^+ + e^-$

(h) $n + p \rightarrow \pi^0 + \pi^+$

(i) $\mu^+ + \mu^- \rightarrow \nu_e + \bar{\nu}_e$

(j) $\pi^- + p \rightarrow K^- + \Sigma^0$

Grand Unified Theories and the Solar Neutrino Problem

17.1 GRAND UNIFIED THEORIES

The standard model as described in the last two chapters is based on certain fundamental beliefs. These include the following:

1. Hadrons are comprised of quarks.
2. Quark number, lepton number, and lepton generation are conserved.
3. Neutrinos are massless.

There is considerable experimental evidence in favor of the quark model and no convincing experimental observations have provided direct evidence of nonconservation of quark or lepton number. However, results of some recent neutrino experiments have been interpreted as evidence for nonconservation of lepton generation and the existence of massive neutrinos. From a theoretical standpoint, work beyond the standard model has provided a fundamental basis for some observations that are difficult to explain in terms of the above assumptions. This involves a unification of the strong and electroweak interactions along the lines of the unification of weak and electromagnetic interactions that is incorporated into the standard model. The resulting theories are referred to as *grand unified theories* or GUTs. This approach can be viewed phenomonologically in the following way. The electroweak interaction increases in strength with decreasing distance while, for small distances at least, the strong interaction decreases in strength with decreasing distance (recall the arguments for confinement). At some distance the magnitude of the three interactions will become

245

comparable and this defines a mass/energy scale on which the interactions become unified. In the simplest GUTs, it is expected that this unification will take place on a distance scale of about 10^{-31} m and this corresponds to an energy of about 10^{15} GeV. (Current theories suggest that the energy scale for unification may be substantially higher than this estimate.) This energy is 10 or more orders of magnitude greater than that which could be obtained by an accelerator using current technology. At this energy leptons and quarks become equivalent. Interactions can be mediated by a (yet undiscovered) very massive gauge boson (called the X-boson) that can couple quarks and leptons. Thus at this energy the conservation of quark number or the conservation of lepton number becomes irrelevant. While a detailed discussion of GUTs is beyond the scope of the present book, the analysis of some experimental data in the context of GUTs provides some interesting insight.

One important consequence of GUTs concerns the stability of the proton. If quark number and lepton number are not conserved then the proton can decay. It is believed that if this occurs, then the principal mode will be

$$p \rightarrow \pi^0 + e^+, \tag{17.1}$$

as shown in Figure 17.1. The possibility of proton decay has been investigated experimentally using large water-filled detectors. The presence of decay byproducts can be observed with light-sensitive detectors from the Cherenkov radiation they produce. This method, as described further in the next section, is an important method for detecting neutrinos. Proton decay has not been observed and present experimental results place a lower limit on the lifetime of the decay in equation (17.1) of about 3×10^{33} years. This value is inconsistent with the predictions of the simplest GUTs, but may be accommodated by more complex GUTs and further experimental work is continuing.

A number of experiments dealing with the properties of neutrinos are of particular interest when considering the possible validity of GUTs. Some of the more significant results are discussed in the next two sections.

Figure 17.1 | Proton Decay (Equation (17.1)) Mediated by a Virtual X-Boson

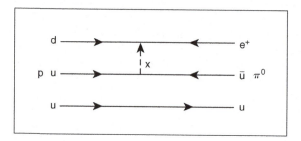

17.2 SOLAR NEUTRINOS

The production of energy in the sun was discussed in Chapter 13. Most of the energy (over 98%) is produced by the proton–proton cycle as illustrated in Figure 13.4. This process produces neutrinos in two kinds of reactions: β^+-decay processes such as

$$p + p \rightarrow d + e^+ + \nu_e \tag{17.2}$$

and electron capture processes such as

$$^7Be + e^- \rightarrow {}^7Li + \nu_e \tag{17.3}$$

In both cases the neutrinos are electron neutrinos and are neutrinos (rather than antineutrinos). In the former processes the neutrinos have a continuous distribution of energies up to the endpoint energy given by the value of Q. In the latter processes the neutrinos have discrete energies. The overall energy spectrum of the neutrinos that are emitted from the sun depends on the branching ratios of the various processes involved. These ratios depend on the internal temperature and composition of the sun. The expected energy spectrum of the neutrinos from the processes shown in Figure 13.4 is illustrated in Figure 17.2.

Figure 17.2 | Expected Neutrino Energy Spectrum Resulting from the Proton–Proton Cycle Shown in Figure 13.4

The neutrinos from electron capture by ^7Be have two discrete energies because the decay can go to either the ground state or an excited state of ^7Li. The threshold energies for the experiments described in the text are indicated.

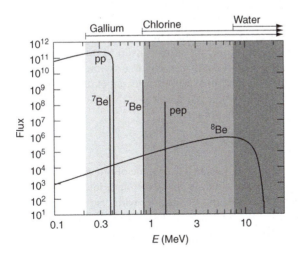

From J. N. Bahcall, *Ap. J.* 467 (1996), 475. Reproduced by permission of the American Astronomical Society.

Table 17.1 | Endpoint Energies, E_{max}, for β-Decay Processes in the Proton–Proton Cycle

Decay	E_{max} (MeV)
$p + p \rightarrow d + e^+ + \nu_e$	0.42
$p + e^- + p \rightarrow d + \nu_e$	1.44
$^7Be + e^- \rightarrow {}^7Li + \nu_e$	0.383, 0.861
$^8B \rightarrow {}^8Be + e^+ + \nu_e$	~15.0

Endpoint energies for these decays are given in Table 17.1. Since the 1960s there has been substantial experimental work on the detection of solar neutrinos. The cross section for such reactions is very small because neutrinos only interact via the weak interaction. As a result, neutrinos are difficult to detect. The most common method of neutrino detection is by the reaction

$$\nu_e + n \rightarrow p + e^-. \tag{17.4}$$

The byproducts of this reaction can be observed by standard charged particle detection methods or by radiochemical methods as described below.

The most extensive solar neutrino studies were initiated in the late 1960s in the Homestake mine in South Dakota where a container of 6.1×10^8 g of C_2Cl_4 (perchlorethylene) is used as a detector. The chlorine consists of about 24% naturally occurring ^{37}Cl. Neutrinos with sufficient energy can produce the reaction

$$\nu_e + {}^{37}Cl \rightarrow {}^{37}Ar + e^-. \tag{17.5}$$

The threshold energy for this process is 0.814 MeV. An inspection of Figure 17.2 and Table 17.1 shows that the most significant contribution to this process will be from the ^8B β-decay neutrinos. The p-e-p process neutrinos and the ^7Be electron capture neutrinos will also contribute, but to a lesser extent. Since the process in equation (17.5) is endothermic, ^{37}Ar is not β-stable and will spontaneously decay back to ^{37}Cl by electron capture:

$$^{37}Ar + e^- \rightarrow {}^{37}Cl + \nu_e. \tag{17.6}$$

The lifetime for this decay is about 50 days. On a time scale of one to two months the argon is removed from the detector fluid by chemical methods. The amount of argon present is determined by observing the decay products from equation (17.6). When electron capture occurs a vacancy is left in one of the inner electron shells of the daughter ^{37}Cl atom. This vacancy is filled by another atomic electron and the difference in binding energies can liberate an outer shell electron. This is the so-called *Auger process* and the liberated electron is an *Auger electron*. The energy spectrum of the Auger electrons is characteristic of the particular decay process and is a convenient means for determining the quantity of

Table 17.2 | Specifications and Results of Solar Neutrino Experiments

Experiment	Detector Nuclei	Mass of Detector Nuclei (g)	Threshold (MeV)	Predicted Flux (SNU)	Measured Flux (SNU)
Homestake	^{37}Cl	1.3×10^8	0.814	8.0 ± 1.0	2.55 ± 0.25
GALLEX	^{71}Ga	1.2×10^7	0.233	132 ± 8	76.2 ± 6.5
SAGE	^{71}Ga	2.3×10^7	0.233	132 ± 8	73 ± 8.5

^{37}Ar atoms present. This, in turn, can be related to the number of neutrinos detected and is a function of (1) the flux of solar neutrinos, (2) the number of ^{37}Cl nuclei in the detector, and (3) the cross section of the reaction in equation (17.5). In the present case the first two factors are large, while the third factor is very small. The number of detected neutrinos is generally expressed in *SNUs* (solar neutrino units). One SNU corresponds to one neutrino detected per 10^{36} detector nuclei per second.

After collecting approximately 30 years of data, the Homestake experiment has measured a solar neutrino flux of 2.55 ± 0.25 SNU. The standard solar model predicts a neutrino flux of 8.0 ± 1.0 SNU for this experiment. Information about the Homestake experiment is summarized in Table 17.2. Clearly the agreement between theory and experiment is less than ideal. There are three possible sources of uncertainty that could account for this discrepancy: (1) the sun does not produce the flux of neutrinos as predicted by the standard solar model, (2) the calibration of the detector is different than expected, or (3) we do not properly understand the behavior of neutrinos. We consider the possibility of these three situations below and in the next section.

A major problem with the Homestake experiment is that it is sensitive to only a small fraction of the total solar neutrino flux. The p–p process is responsible for the majority of solar neutrinos and this flux can be predicted to an accuracy of better than 1% by solar models. The neutrino flux from other solar processes in the proton–proton chain is a very sensitive function of solar model parameters (such as temperature) and can be predicted less accurately. It would, therefore, be highly desirable to undertake an experiment that would be sensitive to at least some of the p–p neutrinos below their 0.42 MeV endpoint energy. A reaction that is useful in this respect is

$$\nu_e + {}^{71}\text{Ga} \rightarrow {}^{71}\text{Ge} + e^-. \tag{17.7}$$

This has a threshold energy of 0.23 MeV and is sensitive to at least some of the neutrinos for each of the proton–proton chain processes. Two experiments along these lines have been undertaken in recent years. One experiment, SAGE, utilizes a detector containing metallic Ga. This is located in Baksan, Russia, and has been in operation since 1990. A second experiment, GALLEX, is located outside of Rome and utilizes a solution of $GaCl_3$ as the detector. This has been collecting data since 1991. In both cases neutrino detection is by radiochemical methods. The resulting ^{71}Ge decays back to ^{71}Ga by β^+-decay with a lifetime of 16.5 days and periodic Ge extraction and analysis provides a measure of the number of neutrinos detected. Information about these two experiments is given in Table 17.2.

The two Ga experiments are consistent with one another but these experimental measurements are not in agreement with the solar model predictions. One concern with the Ga experiments, that is the calibration of the detector, has been dealt with by GALLEX. In 1994 a manufactured ^{51}Cr neutrino source was used to calibrate the sensitivity of the detector. The ^{51}Cr is prepared by neutron irradiation of enriched ^{50}Cr:

$$^{50}Cr + n \rightarrow {}^{51}Cr + \gamma. \qquad (17.8)$$

Neutrinos are produced by the electron capture decay of ^{51}Cr (lifetime 40 days):

$$e^- + {}^{51}Cr \rightarrow {}^{51}V + \nu_e. \qquad (17.9)$$

The strength of the source (which is about 1.7 MCi) is determined by monitoring γ-ray intensity during neutron irradiation and monitoring γ-rays from the ^{51}V excited state deexcitation during the electron capture decay. Experimental results indicate that the measured neutrino flux is $97 \pm 11\%$ of the flux calculated on the basis of the measured source strength. This suggests that for the Ga experiments, at least, errors in detector calibration cannot account for the differences between measurement and theory.

None of the experiments described above is a real-time experiment, nor is any sensitive to the direction of the neutrino flux. A new group of experiments to detect neutrinos deals with these factors. The Kamiokande experiment, located in Japan, served as a neutrino detector from 1986 to 1996 and has since been replaced by a larger version, Super-Kamiokande. These are water-filled detectors that detect neutrino–electron scattering:

$$\nu_e + e^- \rightarrow \nu_e + e^-. \qquad (17.10)$$

Electrons, liberated by their interaction with neutrinos, travel through the water at high energy. The Cherenkov radiation that is produced is detected by a large number of photomultiplier tubes located within the detector. The threshold energy for neutrino detection is about 7 MeV. Thus only the high energy ^{8}B solar neutrinos can be detected. Since the direction of the electron's trajectory is the same as that of the incident neutrino, the origin of the neutrino flux can be investigated. Figure 17.3 shows the results of a measurement of the neutrino flux as a function of angle measured with respect to the direction to the sun. The increase in flux in the forward direction is direct evidence that the neutrinos that are detected (or at least some of them) come from the sun. The incident neutrino energy can also be determined from a measurement of the electron energy. Data for the measured neutrino energy spectrum are compared with the predictions of the standard solar model in Figure 17.4. Overall the measured solar neutrino flux is about half of that predicted by the solar model.

Figure 17.3 | Neutrino Flux Measured by the Super-Kamiokande Detector as a Function of Angle Relative to the Direction to the Sun

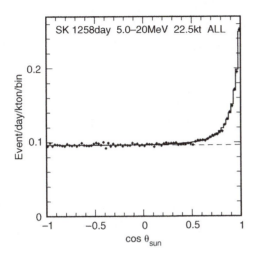

Figure courtesy of the Super-Kamiokande Collaboration.

Figure 17.4 | Neutrino Energy Spectrum Obtained by Super-Kamiokande; Experimental Data (Circles) and Solar Model Predictions (Solid Lines)

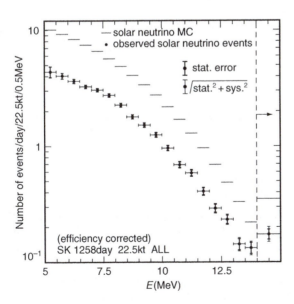

Figure courtesy of the Super-Kamiokande Collaboration.

The results described above indicate that all solar neutrino experiments to date measure a flux that is less than that predicted by the standard solar model. In the following section we will consider the possible reasons for this discrepancy.

17.3 NEUTRINO OSCILLATIONS

The different solar neutrino experiments described above are sensitive to neutrinos from different solar processes. Although all experiments are inconsistent with the predictions of the standard solar model we have not considered whether these different experiments are consistent with one another. Before we discuss this point in detail let's look at the possibilities for the discrepancy between the experimental and model results. In the case of the gallium experiments, careful calibration of the detector seems to indicate that the measurement is indeed a true representation of the actual neutrino flux. This means that the possible reason for the discrepancy is either an incorrect assumption in the solar model or our lack of understanding of the fundamental properties of the neutrino.

The flux of solar neutrinos at higher energy (for example, primarily the ^{8}B neutrinos) is relatively sensitive to minor changes in the parameters of the standard solar model such as the internal temperature of the sun. As well, the cross sections for these processes are not known to a high degree of accuracy. However, the flux of p–p neutrinos is believed to be known to better than 1% on the basis of solar models and the gallium experiments are sensitive to these neutrinos. There is just not enough uncertainty in the standard solar model to account for the observed discrepancies in all the experiments.

The fundamental behavior of neutrinos is a complex problem. Of importance is an understanding of the distinction between the different flavors of neutrino and the distinction between neutrinos and antineutrinos. We expect that neutrinos produced by β^+-decay:

$$p \rightarrow n + e^+ + \nu_e \qquad (17.11)$$

to produce the reaction

$$\nu_e + n \rightarrow p + e^- \qquad (17.12)$$

when they are incident upon neutrons. However, we would not expect neutrinos from the decay of a positive pion to produce the reaction:

$$\nu_\mu + n \rightarrow p + e^-, \qquad (17.13)$$

as this would violate lepton generation number conservation. Over the past 30 years or more a number of experiments have looked for processes of the type of equation (17.13). To date there is no convincing evidence that such processes occur.

A second consideration deals with the distinction between neutrinos and antineutrinos. If neutrinos and antineutrinos (of the same generation) are distinct particles, then we would not expect to observe processes such as

$$\nu_e + p \rightarrow n + e^+ \qquad (17.14)$$

as this would violate conservation of lepton number. If neutrinos and antineutrinos are not distinct particles (as is the case for some neutral mesons), then processes such as (17.14) could occur. Neutrinos that are the same as their antiparticles are referred to as *Majorana neutrinos* and those that are distinct from their antiparticles are referred to as *Dirac neutrinos*. It is well established that double β-decay of the form

$$2n \rightarrow 2p + 2e^- + 2\nu_e. \tag{17.15}$$

has been observed experimentally. If neutrinos are Majorana neutrinos then equation (17.15) would be equivalent to

$$2n \rightarrow 2p + 2e^-. \tag{17.16}$$

This is referred to as neutrinoless double β-decay and, if it exists, can be distinguished from $2\nu_e$ double β-decay on the basis of the energy spectrum of the emitted electrons. To date there is no convincing experimental evidence that neutrinoless double β-decay occurs.

The neutrinos that are produced by the sun are electron neutrinos and, if we accept the direct experimental evidence as described above, these are distinct from muon and tau neutrinos. Evidence also shows that these are Dirac neutrinos and that they are distinct from their antiparticles. These observations are consistent with the standard model but do not help to explain the fact that fewer solar neutrinos are observed than are expected. The possibility that neutrinos have mass is an important consideration for the resolution of this problem.

GUTs allow for the possibility that neutrino flavor states are not pure states but are described as admixtures of more than one state. The theoretical development for neutrino flavor state mixing follows along the lines of quark flavor mixing as described by Cabibbo theory in Section 16.3. If we consider the mixing of two neutrino flavor states then, as a function of time, the neutrino will oscillate between the two flavors. As the neutrino propagates through space the wavelength of this oscillation is given by

$$\lambda = \frac{4\pi\hbar p}{\left(m_1^2 - m_2^2\right)c^2} \tag{17.17}$$

where p is the neutrino momentum and the $m_{1,2}$ are the masses of the two states. (These can (for example) refer to the masses of the electron neutrino and muon neutrino.) When the neutrinos interact with nearby electrons, coherent scattering will enhance this effect. This will occur whenever the neutrinos pass through matter and is referred to as the *Mikheyev-Smirnov-Wolfenstein (MSW) effect*. The MSW effect is important for solar neutrinos as they pass through the dense inner portions of the sun before travelling through space to the earth. This is a plausible explanation for the discrepancy between the calculated and measured solar neutrino fluxes, particularly when taken in conjunction with the further evidence as presented in the following material.

In addition to experiments dealing with solar neutrinos, the measurement of the properties of neutrinos that are created in the atmosphere can provide important information. Atmospheric neutrinos are produced at altitudes of a few tens of kilometers when cosmic rays interact with various particles in the earth's atmosphere. They result primarily from the formation and decay of pions:

$$\pi^+ \to \mu^+ + v_\mu. \tag{17.18}$$

This is followed by decay of the muons:

$$\mu^+ \to e^+ + v_e + \overline{v}_\mu. \tag{17.19}$$

Similar process can involve negative pions leading to the ratio:

$$\frac{n(v_\mu) + n(\overline{v}_\mu)}{n(v_e) + n(\overline{v}_e)} = 2. \tag{17.20}$$

The muon neutrinos produced in this manner are high energy neutrinos (>100 MeV) and can be observed by the Super-Kamiokande experiment as they interact via the weak interaction with electrons in the detector:

$$v_\mu + e^- \to v_\mu + e^-. \tag{17.21}$$

This interaction has a much smaller cross section (by about a factor of six) than the reaction of electron neutrinos and electrons in equation (17.10). This can be seen to be the case as the process in equation (17.10) can be mediated by either a Z^0 or $W^\pm$ boson (see Figure 17.5a) while the reaction in equation (17.21) can

Figure 17.5 | (a) Interaction of an Electron and an Electron Neutrino Mediated by Neutral and Charged Weak Bosons and (b) Interaction of an Electron and a Muon Neutrino Mediated by a Neutral Weak Boson

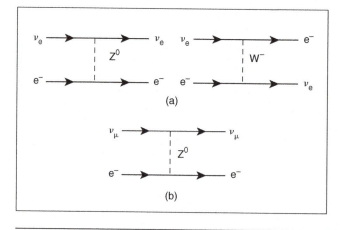

only be mediated by a Z^0 boson (see Figure 17.5b). Since Super-Kamiokande is capable of measuring the direction from which neutrinos have entered the detector, it can distinguish between neutrinos that have been created in the atmosphere directly above the detector (within about 20 km) and those that have been created on the opposite side of the earth (about 13,000 km away). Muon neutrinos that have traveled only about 20 km before being detected do not have time to oscillate into other flavors while those that travel from the other side of the earth may no longer be muon neutrinos when they are incident on the detector. Thus a measure of the muon neutrino flux as a function of angle (relative to the normal to the surface of the earth) can be used to observe neutrino oscillations. Some experimental results are shown in Figure 17.6. The angular dependence of the neutrino flux ratio as seen in the figure is a manifestation of neutrino flavor oscillations. At present, evidence suggests that the muon neutrinos oscillate to become primarily tau neutrinos.

Several new neutrino detectors are planned, are under construction, or, in one case, have recently become operational. The *Sudbury Neutrino Observatory* (SNO) in Ontario, Canada, began taking data in the spring of 1999. This detector is designed along the lines of Kamiokande except that the detector medium is D_2O

Figure 17.6 | Angle Dependence of the Normalized Muon to Electron Neutrino Flux as Observed by Super-Kamiokande

Angles are measured relative to the surface of the earth where $\cos\theta = -1$ represents upward-going neutrinos and $\cos\theta = +1$ represents downward-going neutrinos.

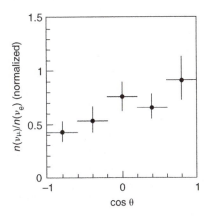

Reprinted from Y. Fukuda *et al.*, *The Super-Kamiokande Collaboration*, "Study of atmospheric neutrino flux in the multi-GeV energy range," *Physics Letters B 436* (1998), 33–41, copyright 1998. Reprinted with permission from Elsevier Science.

rather than H_2O. Three processes are possible:

$$\nu_e + d \rightarrow 2p + e^-, \tag{17.22}$$

$$\nu_x + d \rightarrow \nu_x + p + n, \tag{17.23}$$

and

$$\nu_x + e \rightarrow \nu_x + e. \tag{17.24}$$

The first reaction is sensitive to only electron neutrinos and has a threshold of about 7 MeV. The second and third reactions are flavor insensitive. Thus SNO is sensitive to many of the muon and tau neutrinos that result from the oscillation of solar electron neutrinos, if such oscillations are responsible for the missing solar neutrinos. Preliminary results when combined with the results from Kamiokande indicate that about two-thirds of the electron neutrinos produced by the sun change flavor before they reach the earth. The total flux of all three flavors of neutrinos as detected on earth is in agreement with the electron neutrino flux predicted at the sun by the standard solar model. Further experiments will lead to a better understanding of the previous results of solar and atmospheric neutrino measurements and to a more thorough understanding of the basic properties of neutrinos themselves.

17.4 NEUTRINO MASSES

According to equation (17.17), the observation of neutrino oscillations provides evidence for neutrinos with mass. Although experimental results can provide information about the difference between the square of the mass for the two flavor states, it is important to note that the mixing angle (see Section 16.3) must also be determined in a consistent way. In fact, there is sufficient flexibility in the theoretical predictions concerning masses and mixing angles, that the various experimental results can, for the most part, be viewed in a consistent manner. A more direct method for determining the electron neutrino mass, or at least a means of putting an upper limit on it, is based on a measure of the electron energy spectrum from β-decay. Certain details of the energy spectrum are affected by the presence or absence of neutrino mass. Specifically, changes in the electron energy spectrum will be most significant near the endpoint energy, where the neutrino has its minimum energy. In the case of a massive neutrino the rest mass must be considered even in the absence of kinetic energy. To within the accuracy of the experimental measurements, the analysis of the shape of the energy spectrum near the endpoint energy will allow for a determination of neutrino mass. The most convenient decay process for the study of these effects is

$$^3H \rightarrow {}^3He + e^- + \bar{\nu}_e. \tag{17.25}$$

Figure 17.7 | Energy Spectrum of Electrons from Tritium β-Decay Near the Endpoint Energy and Predictions for Different Electron Neutrino Masses

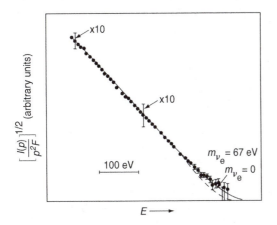

Reprinted from Karl-Erik Bergkvist, "A high-luminosity, high-resolution study of the end point behaviour of the tritium β-spectrum (I). Basic experimental procedure and analysis with regard to neutrino mass and neutrino degeneracy," *Nuclear Physics B 39* (1972), 317–370, copyright 1972. Reprinted with permission from Elsevier Science.

This decay has several desirable features:

1. It is a ground state to ground state transition.
2. The endpoint energy is small (18.6 keV) and this optimizes the effect of neutrino mass on the shape of the energy spectrum.
3. The nuclear eigenstates for the parent and daughter nuclei are well known.

An example of typical experimental results is illustrated in Figure 17.7. In general these experiments do not allow for a precise determination of neutrino mass but allow certain mass limits to be specified. In the case illustrated in the figure the electron neutrino mass seems to be less than about 60 eV, although the question of whether it is identically zero or nonzero cannot be answered on the basis of these data. More recent experiments have pushed the mass limit lower but, again, do not necessarily rule out massless electron neutrinos.

If neutrinos do have mass, then the mass of the muon neutrino will be larger than the mass of the electron neutrino and the mass of the tau neutrino will be larger still. An estimate of the muon and tau neutrino masses can be made on the basis of an evaluation of the masses and energies of the particles involved in certain decay processes. For example, in the decay of a pion:

$$\pi^+ \rightarrow \mu^+ + \nu_\mu, \tag{17.26}$$

the masses of the pion and the muon are known to some degree of accuracy. Measuring the kinetic energies of the pion and muon will allow for limits to be placed on the possible muon neutrino mass. Similarly, the decay of the τ-lepton:

$$\tau^- \rightarrow \mu^- + \nu_\tau + \bar{\nu}_\mu, \tag{17.27}$$

will allow for limits to be placed on the mass of the tau neutrino. Recent studies have indicated the following upper limits on the masses of the three generations of neutrinos:

$$m(\nu_e) < 2.2 \text{ eV}$$

$$m(\nu_\mu) < 170 \text{ keV} \tag{17.28}$$

$$m(\nu_\tau) < 15.5 \text{ MeV}.$$

These results do not, in general, rule out massless neutrinos. However, when combined with the observed neutrino oscillations as described above, they are suggestive of an overall picture consistent with the GUTs approach.

The results presented in this chapter represent only a small fraction of the theoretical and experimental work that has been done on grand unified theories. However, the experimental results as presented here represent some of the most convincing evidence that particle physics is much more complex that the standard model suggests. There is considerable experimental evidence for neutrino oscillations and for a nonzero neutrino mass. This can have profound implications for the overall evolution of the universe and further work, both experimental and theoretical, is necessary in order to fully understand the behavior of neutrinos.

Problems

17.1. Solar neutrinos that are incident on the earth have an average cross section of 10^{-20} b for interaction with a nucleus. Calculate the probability that a neutrino incident on the earth along a line passing through the center of the earth will interact with a nucleus.

17.2. Calculate the threshold energy for electron neutrino absorption by the following nuclei: (a) ^{55}Mn, (b) ^{40}Ca, (c) ^{125}Te, (d) ^{146}Nd, and (e) ^{136}Xe.

Physical Constants and Conversion Factors

Quantity	Symbol	Value	Units
Alpha particle binding energy	B_α	28.296	MeV
Alpha particle mass	m_α	4.00150618	u
		3727.409	MeV/c^2
Atomic mass unit	u	$1.6605402 \times 10^{-27}$	kg
		931.494	MeV/c^2
Avogadro's number	N_A	6.0221367×10^{23}	mole^{-1}
Barn	b	10^{-28}	m^2
		10^2	fm^2
Bohr magneton	μ_B	$9.2740154 \times 10^{-24}$	J $\cdot$ T^{-1}
		5.7883826×10^{-5}	eV $\cdot$ T^{-1}
Bohr radius	a_0	$5.29177249 \times 10^{-11}$	m
Boltzmann's constant	k_B	1.380658×10^{-23}	J $\cdot$ K^{-1}
		8.61738×10^{-5}	eV $\cdot$ K^{-1}
Coulomb constant	$1/(4\pi\varepsilon_0)$	8.987551×10^9	N $\cdot$ m^2 $\cdot$ C^2
	$e^2/(4\pi\varepsilon_0)$	1.439976	MeV $\cdot$ fm
Curie	Ci	3.7×10^{10}	decays $\cdot$ s^{-1}
Deuteron binding energy	B_d	2.224	MeV

(continued)

Quantity	Symbol	Value	Units
Deuteron mass	m_d	2.013553214	u
		1875.628	MeV/c^2
Electron mass	m_e	$5.48579903 \times 10^{-4}$	u
		0.5109988	MeV/c^2
Electron volt	eV	$1.60217733 \times 10^{-19}$	J
Electronic charge	e	$1.60217733 \times 10^{-19}$	C
Fine structure constant	α	0.00729735308	
Gas constant	R	8.314510	$J \cdot K^{-1} \cdot mol^{-1}$
Gravitational constant	G	6.67259×10^{-11}	$N \cdot m^2 \cdot kg^{-2}$
Neutron magnetic moment	μ_n	-1.913	μ_N
Neutron mass	m_n	1.008664904	u
		939.56531	MeV/c^2
Nuclear magneton	μ_N	$5.0507866 \times 10^{-27}$	$J \cdot T^{-1}$
		3.1524517×10^{-8}	$eV \cdot T^{-1}$
Permeability of free space	μ_0	$4\pi \times 10^{-7}$	$T \cdot m \cdot A^{-1}$
Permittivity of free space	ε_0	$8.854187817 \times 10^{-12}$	$C^2 \cdot N^{-1} \cdot m^{-2}$
Planck's constant	h	$6.6260755 \times 10^{-34}$	$J \cdot s$
		4.13570×10^{-15}	$eV \cdot s$
	$\hbar$	$1.05457266 \times 10^{-34}$	$J \cdot s$
		6.58217×10^{-16}	$eV \cdot s$
	hc	1.240×10^3	$MeV \cdot fm$
	$\hbar c$	1.973×10^2	$MeV \cdot fm$
Proton magnetic moment	μ_p	2.793	μ_N
Proton mass	m_p	1.007276470	u
		938.2723	MeV/c^2
Rydberg constant	R_∞	1.0973731534×10^7	m^{-1}
Speed of light	c	2.99792458×10^8	$m \cdot s^{-1}$
Stefan-Boltzmann constant	σ	5.67051×10^{-8}	$W \cdot m^{-2} \cdot K^{-4}$

Properties of Nuclides

The following table gives ground state properties of selected nuclides. Spin and parity states that are not known with certainty are indicated in parentheses. Natural abundances are shown for stable and naturally occurring nuclides.

Z		A	Mass (u)	J^π	Abundance or Half-life	Z		A	Mass (u)	J^π	Abundance or Half-life
1	H	1	1.007825032	$\frac{1}{2}^+$	99.985%	5	B	8	8.024606713	2^+	770ms
		2	2.014101778	1^+	0.015%			10	10.01293703	3^+	19.9%
		3	3.016049268	$\frac{1}{2}^+$	12.33y			11	11.00930547	$\frac{3}{2}^-$	80.1%
2	He	3	3.01602931	$\frac{1}{2}^+$	0.00014%			12	12.01435211	1^+	20.20ms
		4	4.00260325	0^+	99.99986%			13	13.01778027	$\frac{3}{2}^-$	17.36ms
		5	5.012223628	$\frac{3}{2}^+$				14	14.02540406	2^-	13.8ms
		6	6.018888072	0^+	806.7ms	6	C	9	9.031040087	$(\frac{3}{2}^-)$	126.5ms
3	Li	6	6.015122281	1^+	7.5%			10	10.01685311	0^+	19.255s
		7	7.016004049	$\frac{3}{2}^-$	92.5%			11	11.01143382	$\frac{3}{2}^-$	20.39m
		8	8.02248667	2^+	838ms			12	12.00000000	0^+	98.9%
		9	9.026789122	$\frac{3}{2}^-$	178.3ms			13	13.00335484	$\frac{1}{2}^-$	1.1%
4	Be	6	6.019725804	0^+				14	14.00324199	0^+	5730y
		7	7.016929246	$\frac{3}{2}^-$	53.29d			15	15.01059926	$\frac{1}{2}^+$	2.449s
		8	8.00530594	0^+	0.07fs			16	16.01470124	0^+	0.747s
		9	9.012182135	$\frac{3}{2}^-$	100%	7	N	12	12.0186132	1^+	11.000ms
		10	10.01353372	0^+	1.51My			13	13.00573858	$\frac{1}{2}^-$	9.965m
		11	11.02165765	$\frac{1}{2}^+$	13.81s			14	14.00307401	1^+	99.63%
		12	12.02692063	0^+	23.6ms			15	15.0001089	$\frac{1}{2}^-$	0.37%

(continued)

Z	A	Mass (u)	J^π	Abundance or Half-life	Z	A	Mass (u)	J^π	Abundance or Half-life
	16	16.00610142	2^-	7.13s		29	28.98855474	$\frac{3}{2}^+$	1.30s
	17	17.00844967	$\frac{1}{2}^-$	4.173s		30	29.99046453	0^+	335ms
	18	18.01408183	1^-	624ms	13 Al	24	23.99994091	4^+	2.053s
8 O	13	13.0248104	$(\frac{3}{2}^-)$	8.58ms		25	24.99042856	$\frac{5}{2}^+$	7.183s
	14	14.00859529	0^+	70.606s		26	25.98689166	5^+	0.74My
	15	15.00306539	$\frac{1}{2}^-$	122.24s		27	26.98153844	$\frac{5}{2}^+$	100%
	16	15.99491462	0^+	99.76%		28	27.98191018	3^+	2.2414m
	17	16.9991315	$\frac{5}{2}^+$	0.038%		29	28.98044485	$\frac{5}{2}^+$	6.56m
	18	17.99916042	0^+	0.2%		30	29.9829603	3^+	3.60s
	19	19.00357873	$\frac{5}{2}^+$	26.91s		31	30.98394602	$(\frac{3}{2},\frac{5}{2})^+$	644ms
	20	20.00407615	0^+	13.51s		32	31.98812438	1^+	33ms
	21	21.00865463	$(\frac{1}{2},\frac{3}{2},\frac{5}{2})^+$	3.42s	14 Si	24	24.01154571	0^+	102ms
	22	22.00996716	0^+	2.25s		25	25.00410664	$\frac{5}{2}^+$	220ms
9 F	17	17.00209524	$\frac{5}{2}^+$	64.49s		26	25.99232994	0^+	2.234s
	18	18.00093767	1^+	109.77m		27	26.98670476	$\frac{5}{2}^+$	4.16s
	19	18.99840321	$\frac{1}{2}^+$	100%		28	27.97692653	0^+	92.23%
	20	19.99998132	2^+	11.00s		29	28.97649472	$\frac{1}{2}^+$	4.67%
	21	20.99994892	$\frac{5}{2}^+$	4.158s		30	29.97377022	0^+	3.1%
	22	22.00299925	$4^+, (3^+)$	4.23s		31	30.97536328	$\frac{3}{2}^+$	157.3m
	23	23.00357439	$(\frac{3}{2},\frac{5}{2})^+$	2.23s		32	31.97414813	0^+	172y
	24	24.00809937	$(1, 2, 3)^+$	0.34s		33	32.97800052	$(\frac{3}{2}^+)$	6.18s
10 Ne	17	17.01769757	$\frac{1}{2}^-$	109.2ms		34	33.97857575	0^+	2.77s
	18	18.00569707	0^+	1672ms	15 P	28	27.99231233	3^+	270.3ms
	19	19.00187984	$\frac{1}{2}^+$	17.34s		29	28.98180138	$\frac{1}{2}^+$	4.142s
	20	19.99244018	0^+	90.48%		30	29.97831381	1^+	2.498m
	21	20.99384674	$\frac{3}{2}^+$	0.27%		31	30.97376151	$\frac{1}{2}^+$	100%
	22	21.99138551	0^+	9.25%		32	31.97390716	1^+	14.262d
	23	22.99446734	$\frac{5}{2}^+$	37.24s		33	32.97172528	$\frac{1}{2}^+$	25.34d
	24	23.99361507	0^+	3.38m		34	33.97363638	1^+	12.43s
	25	24.9977899	$(\frac{1}{2},\frac{3}{2})^+$	602ms		35	34.97331425	$\frac{1}{2}^+$	47.3s
	26	26.0004615	0^+	0.23s	16 S	30	29.98490295	0^+	1.178s
11 Na	20	20.00734826	2^+	447.9ms		31	30.97955442	$\frac{1}{2}^+$	2.572s
	21	20.9976551	$\frac{3}{2}^+$	22.49s		32	31.97207069	0^+	95.02%
	22	21.99443678	3^+	2.6019y		33	32.9714585	$\frac{3}{2}^+$	0.75%
	23	22.98976968	$\frac{3}{2}^+$	100%		34	33.96786683	0^+	4.21%
	24	23.99096333	4^+	14.9590h		35	34.96903214	$\frac{3}{2}^+$	87.51d
	25	24.98995435	$\frac{5}{2}^+$	59.1s		36	35.96708088	0^+	0.02%
	26	25.9925899	3^+	1.072s		37	36.97112572	$\frac{7}{2}^-$	5.05m
	27	26.9940087	$\frac{5}{2}^+$	301ms		38	37.97116344	0^+	170.3m
	28	27.99889041	1^+	30.5ms		39	38.97513528	$(\frac{3}{2},\frac{5}{2},\frac{7}{2})^-$	11.5s
12 Mg	20	20.01886274	0^+	95ms		40	39.97547	0^+	8.8s
	21	21.01171417	$(\frac{3}{2},\frac{5}{2})^+$	122ms	17 Cl	32	31.98568891	1^+	298ms
	22	21.99957406	0^+	3.857s		33	32.9774518	$\frac{3}{2}^+$	2.511s
	23	22.99412485	$\frac{3}{2}^+$	11.317s		34	33.97376197	0^+	1.5264s
	24	23.9850419	0^+	78.89%		35	34.96885271	$\frac{3}{2}^+$	75.77%
	25	24.98583702	$\frac{5}{2}^+$	10%		36	35.96830695	2^+	0.301My
	26	25.98259304	0^+	11.01%		37	36.9659026	$\frac{3}{2}^+$	24.23%
	27	26.98434074	$\frac{1}{2}^+$	9.458m		38	37.96801055	2^-	37.24m
	28	27.9838767	0^+	20.91h		39	38.96800768	$\frac{3}{2}^+$	55.6m

Appendix B: Properties of Nuclides   263

Z	A	Mass (u)	J^π	Abundance or Half-life	Z	A	Mass (u)	J^π	Abundance or Half-life
	40	39.97041556	2^-	1.35m		43	42.96115098	$\frac{7}{2}^-$	3.891h
	41	40.97065021	$(\frac{1}{2},\frac{3}{2})^+$	38.4s		44	43.95940305	2^+	3.927h
18 Ar	32	31.99766066	0^+	98ms		45	44.95591024	$\frac{7}{2}^-$	100%
	33	32.98992872	$\frac{1}{2}^+$	173.0ms		46	45.95517025	4^+	83.79d
	34	33.98027012	0^+	844.5ms		47	46.95240803	$\frac{7}{2}^-$	3.345d
	35	34.97525673	$\frac{3}{2}^+$	1.775s		48	47.95223499	6^+	43.67h
	36	35.96754628	0^+	0.337%		49	48.95002407	$\frac{7}{2}^-$	57.2m
	37	36.96677591	$\frac{3}{2}^+$	35.04d		50	49.95218701	5^+	102.5s
	38	37.96273216	0^+	0.063%		51	50.9536027	$(\frac{7}{2})^+$	12.4s
	39	38.96431341	$\frac{7}{2}^-$	269y		52	51.95665	3^+	8.2s
	40	39.96238312	0^+	99.6%	22 Ti	40	39.99049891	0^+	50ms
	41	40.96450083	$\frac{7}{2}^-$	109.34m		41	40.983131	$\frac{3}{2}^+$	80ms
	42	41.96304639	0^+	32.9y		42	41.97303162	0^+	199ms
	43	42.9656707	$(\frac{3}{2},\frac{5}{2})$	5.37m		43	42.96852334	$\frac{7}{2}^-$	509ms
	44	43.96536527	0^+	11.87m		44	43.95969024	0^+	49y
	46	45.96809347	0^+	8.4s		45	44.95812435	$\frac{7}{2}^-$	184.8m
19 K	35	34.98801162	$\frac{3}{2}^+$	190ms		46	45.95262949	0^+	8%
	36	35.98129341	2^+	342ms		47	46.95176379	$\frac{5}{2}^-$	7.3%
	37	36.97337692	$\frac{3}{2}^+$	1.226s		48	47.94794705	0^+	73.8%
	38	37.96908011	3^+	7.636m		49	48.94787079	$\frac{7}{2}^-$	5.5%
	39	38.96370686	$\frac{3}{2}^+$	93.2581%		50	49.94479207	0^+	5.4%
	40	39.96399867	4^-	0.0117%		51	50.94661602	$\frac{3}{2}^-$	5.76m
	41	40.96182597	$\frac{3}{2}^+$	6.7302%		52	51.94689818	0^+	1.7m
	42	41.96240306	2^-	12.360h		53	52.94973171	$(\frac{3}{2})^-$	32.7s
	43	42.96071575	$\frac{3}{2}^+$	22.3h	23 V	45	44.96578229	$\frac{7}{2}^-$	547ms
	44	43.96155615	2^-	22.13m		46	45.96019949	0^+	422.37ms
	45	44.96069966	$\frac{3}{2}^+$	17.3m		47	46.95490692	$\frac{3}{2}^-$	32.6m
	46	45.9619762	(2^-)	105s		48	47.95225448	4^+	15.9735d
	47	46.961677	$\frac{1}{2}^+$	17.5s		49	48.94851691	$\frac{7}{2}^-$	330d
	48	47.96551295	(2^-)	6.8s		50	49.94716279	6^+	0.25%
	51	50.97638	$(\frac{1}{2}^+,\frac{3}{2}^+)$	365ms		51	50.94396368	$\frac{7}{2}^-$	99.75%
20 Ca	36	35.99308723	0^+	100ms		52	51.94477966	3^+	3.743m
	37	36.98587151	$\frac{3}{2}^+$	175ms		53	52.94434252	$\frac{7}{2}^-$	1.61m
	38	37.97631864	0^+	440ms		54	53.94644438	3^+	49.8s
	39	38.97071773	$\frac{3}{2}^+$	859.6ms		55	54.94723819	$(\frac{7}{2}^-)$	6.54s
	40	39.96259116	0^+	96.941%	24 Cr	46	45.96836165	0^+	0.26s
	41	40.96227835	$\frac{7}{2}^-$	0.103My		47	46.96290651	$\frac{3}{2}^-$	508ms
	42	41.95861834	0^+	0.647%		48	47.95403586	0^+	21.56h
	43	42.95876683	$\frac{7}{2}^-$	0.135%		49	48.95134114	$\frac{5}{2}^-$	42.3m
	44	43.95548109	0^+	2.086%		50	49.94604961	0^+	4.34500%
	45	44.95618594	$\frac{7}{2}^-$	163.8d		51	50.94477177	$\frac{7}{2}^-$	27.702d
	46	45.95369276	0^+	0.004%		52	51.94051190	0^+	83.79000%
	47	46.95454646	$\frac{7}{2}^-$	4.536d		53	52.94065378	$\frac{3}{2}^-$	9.50000%
	48	47.95253351	0^+	0.187%		54	53.93888492	0^+	2.36500%
	49	48.9556733	$\frac{3}{2}^-$	8.715m		55	54.94084416	$\frac{3}{2}^-$	3.497m
	50	49.95751829	0^+	13.9s		56	55.94064524	0^+	5.94m
21 Sc	40	39.97796401	4^-	182.3ms	25 Mn	49	48.95962342	$\frac{5}{2}^-$	384ms
	41	40.96925132	$\frac{7}{2}^-$	596.3ms		50	49.95424396	0^+	283.07ms
	42	41.96551676	0^+	681.3ms					

(continued)

Z	A	Mass (u)	J^π	Abundance or Half-life
	51	50.94821549	$\frac{5}{2}^-$	46.2m
	52	51.94557008	6^+	5.591d
	53	52.9412947	$\frac{7}{2}^-$	3.74My
	54	53.94036325	3^+	312.3d
	55	54.93804964	$\frac{5}{2}^-$	100%
	56	55.93890937	3^+	2.5785h
	57	56.93828746	$\frac{5}{2}^-$	85.4s
	58	57.93998645	3^+	65.3s
	59	58.94044717	$\frac{3}{2}^-,\frac{5}{2}^-$	4.6s
26 Fe	51	50.95682494	$(\frac{5}{2}^-)$	305ms
	52	51.94811653	0^+	8.275h
	53	52.94531228	$\frac{7}{2}^-$	8.51m
	54	53.93961484	0^+	5.9%
	55	54.93829803	$\frac{3}{2}^-$	2.73y
	56	55.93494213	0^+	91.72%
	57	56.93539871	$\frac{1}{2}^-$	2.1%
	58	57.93328046	0^+	0.28%
	59	58.93488049	$\frac{3}{2}^-$	44.503d
	60	59.93407694	0^+	1.5My
	61	60.9367946	$\frac{3}{2}^-,\frac{5}{2}^-$	5.98m
	62	61.9367705	0^+	68s
27 Co	53	52.95422499	$(\frac{7}{2}^-)$	240ms
	54	53.94846415	0^+	193.23ms
	55	54.94200315	$\frac{7}{2}^-$	17.53h
	56	55.93984394	4^+	77.27d
	57	56.93629624	$\frac{7}{2}^-$	271.79d
	58	57.93575757	2^+	70.82d
	59	58.93320019	$\frac{7}{2}^-$	100%
	60	59.9338222	5^+	5.2714y
	61	60.93247938	$\frac{7}{2}^-$	1.650h
	62	61.93405421	2^+	1.50m
	63	62.93361522	$(\frac{7}{2})^-$	27.4s
	64	63.93581352	1^+	0.30s
	65	64.93648458	$(\frac{7}{2})^-$	1.20s
28 Ni	53	52.96846	$(\frac{7}{2}^-)$	45ms
	54	53.95791051	0^+	140ms
	55	54.95133633	$\frac{7}{2}^-$	212.1ms
	56	55.94213634	0^+	5.9d
	57	56.93980049	$\frac{3}{2}^-$	35.60h
	58	57.93534792	0^+	68.077%
	59	58.93435155	$\frac{3}{2}^-$	0.076My
	60	59.93079063	0^+	26.223%
	61	60.93106044	$\frac{3}{2}^-$	1.14%
	62	61.92834876	0^+	3.634%
	63	62.92967295	$\frac{1}{2}^-$	100.1y
	64	63.92796957	0^+	0.926%
	65	64.93008801	$\frac{5}{2}^-$	2.5172h
	66	65.92911523	0^+	54.6h
	67	66.93156964	$(\frac{1}{2}^-)$	21s
29 Cu	57	56.9492157	$\frac{3}{2}^-$	199.4ms
	58	57.94454073	1^+	3.204s

Z	A	Mass (u)	J^π	Abundance or Half-life
	59	58.93950411	$\frac{3}{2}^-$	81.5s
	60	59.93736812	2^+	23.7m
	61	60.93346218	$\frac{3}{2}^-$	3.333h
	62	61.9325873	1^+	9.74m
	63	62.92960108	$\frac{3}{2}^-$	69.17%
	64	63.92976787	1^+	12.700h
	65	64.92779371	$\frac{3}{2}^-$	30.83%
	66	65.92887304	1^+	5.088m
	67	66.92775029	$\frac{3}{2}^-$	61.83h
	68	67.929620	1^+	31s
	69	68.92942528	$\frac{3}{2}^-$	2.85m
	70	69.93240929	1^+	4.5s
30 Zn	59	58.94926707	$\frac{3}{2}^-$	182.0ms
	60	59.94183203	0^+	2.38m
	61	60.93951391	$\frac{3}{2}^-$	89.1s
	62	61.93433413	0^+	9.186h
	63	62.93321556	$\frac{3}{2}^-$	38.47m
	64	63.92914658	0^+	48.6%
	65	64.92924508	$\frac{5}{2}^-$	244.26d
	66	65.92603676	0^+	27.9%
	67	66.92713086	$\frac{5}{2}^-$	4.1%
	68	67.92484757	0^+	18.8%
	69	68.92655354	$\frac{1}{2}^-$	56.4m
	70	69.92532487	0^+	0.6%
	71	70.9277272	$\frac{1}{2}^-$	2.45m
	72	71.92686112	0^+	46.5h
	73	72.92977947	$(\frac{1}{2})^-$	23.5s
	74	73.92945826	0^+	96s
	75	74.93293738	$(\frac{7}{2}^+)$	10.2s
31 Ga	62	61.94417961	0^+	116.12ms
	63	62.93914153	$\frac{3}{2}^-,\frac{5}{2}^-$	32.4s
	64	63.93683831	0^+	2.630m
	65	64.93273932	$\frac{3}{2}^-$	15.2m
	66	65.93159236	0^+	9.49h
	67	66.92820492	$\frac{3}{2}^-$	3.2612d
	68	67.9279835	1^+	67.629m
	69	68.92558091	$\frac{3}{2}^-$	60.108%
	70	69.92602774	1^+	21.14m
	71	70.92470501	$\frac{3}{2}^-$	39.892%
	72	71.92636935	3^-	14.10h
	73	72.92516983	$\frac{3}{2}^-$	4.86h
	74	73.926941	(3^-)	8.12m
	75	74.92650065	$\frac{3}{2}^-$	126s
	76	75.92892826	(3^-)	32.6s
32 Ge	61	60.96379	$(\frac{3}{2}^-)$	40ms
	64	63.94157264	0^+	63.7s
	66	65.9338468	0^+	2.26h
	67	66.93273842	$\frac{1}{2}^-$	18.9m
	68	67.92809727	0^+	270.82d
	69	68.927972	$\frac{5}{2}^-$	39.05h
	70	69.92425037	0^+	21.23%

Z	A	Mass (u)	J^π	Abundance or Half-life	Z	A	Mass (u)	J^π	Abundance or Half-life
	71	70.92495399	$\frac{1}{2}^-$	11.43d		85	84.91560803	$\frac{3}{2}^-$	2.90m
	72	71.92207618	0^+	27.66%		87	86.92071071	$\frac{3}{2}^-$	55.60s
	73	72.92345936	$\frac{9}{2}^+$	7.73%	36 Kr	73	72.93893112	$\frac{5}{2}^-$	27.0s
	74	73.92117821	0^+	35.94%		74	73.93325823	0^+	11.50m
	75	74.92285949	$\frac{1}{2}^-$	82.78m		75	74.93103379	$(\frac{5}{2})^+$	4.3m
	76	75.92140272	0^+	7.44%		76	75.9259483	0^+	14.8h
	77	76.92354846	$\frac{7}{2}^+$	11.30h		77	76.92466788	$\frac{5}{2}^+$	74.4m
	78	77.92285289	0^+	88.0m		78	77.92038627	0^+	0.35%
	79	78.92540156	$(\frac{1}{2})^-$	18.98s		79	78.92008299	$\frac{1}{2}^-$	35.04h
	80	79.92544476	0^+	29.5s		80	79.91637804	0^+	2.25%
33 As	67	66.93919042	$(\frac{5}{2}^-)$	42.5s		81	80.91659242	$\frac{7}{2}^+$	0.229My
	69	68.93228015	$\frac{5}{2}^-$	15.2m		82	81.9134846	0^+	11.6%
	70	69.93092781	$4^{(+)}$	52.6m		83	82.91413595	$\frac{9}{2}^+$	11.5%
	71	70.92711472	$\frac{5}{2}^-$	65.28h		84	83.91150663	0^+	57%
	72	71.92675265	2^-	26.0h		85	84.91252695	$\frac{9}{2}^+$	10.756y
	73	72.92382529	$\frac{3}{2}^-$	80.30d		86	85.91061031	0^+	17.3%
	74	73.92392908	2^-	17.77d		87	86.91335425	$\frac{5}{2}^+$	76.3m
	75	74.92159642	$\frac{3}{2}^-$	100%		88	87.91444695	0^+	2.84h
	76	75.92239393	2^-	26.32h		89	88.91763251	$(\frac{3}{2}^+,\frac{5}{2}^+)$	3.15m
	77	76.9206477	$\frac{3}{2}^-$	38.83h		90	89.9195238	0^+	32.32s
	78	77.92182858	2^-	90.7m	37 Rb	79	78.92399672	$\frac{5}{2}^+$	22.9m
	79	78.9209485	$\frac{3}{2}^-$	9.01m		80	79.92251932	1^+	34s
	80	79.92257816	1^+	15.2s		81	80.91899417	$\frac{3}{2}^-$	4.576h
	81	80.92213288	$\frac{3}{2}^-$	33.3s		82	81.91820769	1^+	1.273m
34 Se	70	69.933504	0^+	41.1m		83	82.91511195	$\frac{5}{2}^-$	86.2d
	71	70.932268	$\frac{5}{2}^-$	4.74m		84	83.91438468	2^-	32.77d
	72	71.92711231	0^+	8.40d		85	84.91178934	$\frac{5}{2}^-$	72.17%
	73	72.9267668	$\frac{9}{2}^+$	7.15h		86	85.91116708	2^-	18.631d
	74	73.92247656	0^+	0.89%		87	86.90918347	$\frac{3}{2}^-$	27.83%
	75	74.92252357	$\frac{5}{2}^+$	119.779d		88	87.91131856	2^-	17.78m
	76	75.91921411	0^+	9.36%		90	89.91480894	0^-	158s
	77	76.91991461	$\frac{1}{2}^-$	7.63%		91	90.91653416	$\frac{3}{2}^{(-)}$	58.4s
	78	77.91730952	0^+	23.78%		92	91.91972544	0^-	4.492s
	79	78.9184998	$\frac{7}{2}^+$	0.65My		93	92.92203277	$\frac{5}{2}^-$	5.84s
	80	79.91652183	0^+	49.61%	38 Sr	79	78.92970708	$\frac{3}{2}^{(-)}$	2.25m
	81	80.91799293	$\frac{1}{2}^-$	18.45m		80	79.92452459	0^+	106.3m
	82	81.9167	0^+	8.73%		81	80.9232131	$\frac{1}{2}^-$	22.3m
	83	82.91911907	$\frac{9}{2}^+$	22.3m		82	81.91840126	0^+	25.55d
	84	83.91846452	0^+	3.1m		83	82.91755503	$\frac{7}{2}^+$	32.41h
	87	86.92852075	$\frac{5}{2}^+$	5.85s		84	83.91342478	0^+	0.56%
35 Br	75	74.92577641	$\frac{3}{2}^-$	96.7m		85	84.91293269	$\frac{9}{2}^+$	64.84d
	76	75.92454197	1^-	16.2h		86	85.90926235	0^+	9.86%
	77	76.92138012	$\frac{3}{2}^-$	57.036h		87	86.90887932	$\frac{9}{2}^+$	7%
	78	77.92114613	1^+	6.46m		88	87.90561434	0^+	82.58%
	79	78.91833765	$\frac{3}{2}^-$	50.69%		89	88.90745291	$\frac{5}{2}^+$	50.53d
	80	79.91852995	1^+	17.68m		90	89.9077376	0^+	28.78y
	81	80.91629106	$\frac{3}{2}^-$	49.31%		91	90.91020985	$\frac{5}{2}^+$	9.63h
	82	81.91680467	5^-	35.30h		92	91.9110299	0^+	2.71h
	83	82.91518022	$\frac{3}{2}^-$	2.40h		93	92.91402241	$\frac{5}{2}^+$	7.423m
	84	83.916503	2^-	31.8m		94	93.91535986	0^+	75.3s

(continued)

Z	A	Mass (u)	J^π	Abundance or Half-life	Z	A	Mass (u)	J^π	Abundance or Half-life
	95	94.91935821	$\frac{1}{2}^+$	23.90s		92	91.90681048	0^+	14.84%
	96	95.92168047	0^+	1.07s		93	92.90681221	$\frac{5}{2}^+$	0.0040My
39 Y	83	82.92235257	$(\frac{9}{2}^+)$	7.08m		94	93.90508758	0^+	9.25%
	84	83.92038777	1^+	4.6s		95	94.90584149	$\frac{5}{2}^+$	15.92%
	85	84.91642708	$(\frac{1}{2})^-$	2.68h		96	95.9046789	0^+	16.68%
	86	85.91488772	4^-	14.74h		97	96.90602103	$\frac{5}{2}^+$	9.55%
	87	86.91087783	$\frac{1}{2}^-$	79.8h		98	97.90540785	0^+	24.13%
	88	87.90950336	4^-	106.65d		99	98.9077116	$\frac{1}{2}^+$	65.94h
	89	88.9058479	$\frac{1}{2}^-$	100%		100	99.90747715	0^+	9.63%
	90	89.90715144	2^-	64.10h		101	100.9103465	$\frac{1}{2}^+$	14.61m
	91	90.90730342	$\frac{1}{2}^-$	58.51d		102	101.9102972	0^+	11.3m
	92	91.90894683	2^-	3.54h		103	102.9132046	$(\frac{3}{2}^+)$	67.5s
	93	92.90958158	$\frac{1}{2}^-$	10.18h		104	103.9137584	0^+	60s
	94	93.91159401	2^-	18.7m	43 Tc	93	92.91024847	$\frac{9}{2}^+$	2.75h
	95	94.91282371	$\frac{1}{2}^-$	10.3m		94	93.90965631	7^+	293m
	96	95.91589779	0^-	5.34s		95	94.90765645	$\frac{9}{2}^+$	20.0h
	97	96.91813102	$(\frac{1}{2}^-)$	3.75s		96	95.9078708	7^+	4.28d
40 Zr	84	83.92325	0^+	25.9m		97	96.90636484	$\frac{9}{2}^+$	2.6My
	85	84.92146522	$\frac{7}{2}^+$	7.86m		98	97.90721569	$(6)^+$	4.2My
	86	85.91647285	0^+	16.5h		99	98.90625455	$\frac{9}{2}^+$	0.2111My
	87	86.91481658	$(\frac{9}{2})^+$	1.68h		100	99.90765759	1^+	15.8s
	88	87.91022618	0^+	83.4d		101	100.9073144	$(\frac{9}{2}^+)$	14.22m
	89	88.90888892	$\frac{9}{2}^+$	78.41h		102	101.9092129	1^+	5.28s
	90	89.90470368	0^+	51.45%		103	102.9091788	$\frac{5}{2}^+$	54.2s
	91	90.90564497	$\frac{5}{2}^+$	11.22%	44 Ru	92	91.92012	0^+	3.65m
	92	91.90504011	0^+	17.15%		93	92.91705152	$(\frac{9}{2}^+)$	59.7s
	93	92.90647563	$\frac{5}{2}^+$	1.53My		94	93.91135957	0^+	51.8m
	94	93.90631577	0^+	17.38%		95	94.91041273	$\frac{5}{2}^+$	1.643h
	95	94.90804274	$\frac{5}{2}^+$	64.02d		96	95.90759768	0^+	5.54%
	96	95.90827568	0^+	2.8%		97	96.90755455	$\frac{5}{2}^+$	2.9d
	97	96.91095072	$\frac{1}{2}^+$	16.90h		98	97.90528711	0^+	1.86%
	98	97.91274637	0^+	30.7s		99	98.90593931	$\frac{5}{2}^+$	12.7%
	99	98.91651108	$(\frac{1}{2}^+)$	2.1s		100	99.90421966	0^+	12.6%
	100	99.9177617	0^+	7.1s		101	100.9055822	$\frac{5}{2}^+$	17.1%
41 Nb	87	86.92036144	$(\frac{9}{2}^+)$	2.6m		102	101.9043495	0^+	31.6%
	88	87.917956	(8^+)	14.5m		103	102.9063237	$\frac{3}{2}^+$	39.26d
	89	88.9134955	$(\frac{9}{2}^+)$	1.9h		104	103.9054301	0^+	18.6%
	90	89.91126411	8^+	14.60h		105	104.9077503	$\frac{3}{2}^+$	4.44h
	91	90.90699054	$\frac{9}{2}^+$	680y		106	105.9073269	0^+	373.59d
	92	91.90719321	$(7)^+$	34.7My		107	106.9099072	$(\frac{5}{2})^+$	3.75m
	93	92.90637754	$\frac{9}{2}^+$	100%		108	107.9101922	0^+	4.55m
	94	93.90728346	$(6)^+$	0.0203My		109	108.9132016	$(\frac{5}{2}^+)$	34.5s
	95	94.90683518	$\frac{9}{2}^+$	34.975d	45 Rh	97	96.91133664	$\frac{9}{2}^+$	30.7m
	96	95.90810008	6^+	23.35h		98	97.91071643	$(2)^+$	8.7m
	97	96.90809714	$\frac{9}{2}^+$	72.1m		99	98.9081321	$\frac{1}{2}^-$	16.1d
	98	97.91033069	1^+	2.86s		100	99.90811663	1^-	20.8h
	99	98.91161786	$\frac{9}{2}^+$	15.0s		101	100.9061635	$\frac{1}{2}^-$	3.3y
42 Mo	88	87.92195273	0^+	8.0m		102	101.9068428	$(1^-, 2^-)$	207d
	89	88.91948056	$(\frac{9}{2}^+)$	2.04m		103	102.9055042	$\frac{1}{2}^-$	100%
	90	89.91393616	0^+	5.67h		104	103.9066553	1^+	42.3s
	91	90.91175075	$\frac{9}{2}^+$	15.49m		105	104.9056924	$\frac{7}{2}^+$	35.36h

Z	A	Mass (u)	J^π	Abundance or Half-life	Z	A	Mass (u)	J^π	Abundance or Half-life
	106	105.9072846	1^+	29.80s		115	114.9054306	$\frac{1}{2}^+$	53.46h
	107	106.9067505	$\frac{7}{2}^+$	21.7m		116	115.9047554	0^+	7.49%
	108	107.9087308	(5^+)	6.0m		117	116.9072182	$\frac{1}{2}^+$	2.49h
	109	108.9087356	$\frac{7}{2}^+$	80s		118	117.9069141	0^+	50.3m
46 Pd	97	96.91647892	$(\frac{5}{2}^+)$	3.10m		119	118.9099226	$\frac{3}{2}^+$	2.69m
	98	97.91272075	0^+	17.7m		120	119.9098514	0^+	50.80s
	99	98.91176776	$(\frac{5}{2})^+$	21.4m		121	120.9129804	$(\frac{3}{2}^+)$	13.5s
	100	99.9085046	0^+	3.63d	49 In	106	105.9134611	7^+	6.2m
	101	100.9082891	$(\frac{5}{2}^+)$	8.47h		107	106.9102922	$\frac{9}{2}^+$	32.4m
	102	101.9056077	0^+	1.02%		108	107.9097197	7^+	58.0m
	103	102.9060872	$\frac{5}{2}^+$	16.991d		109	108.9071541	$\frac{9}{2}^+$	4.2h
	104	103.9040349	0^+	11.14%		110	109.9071688	7^+	4.9h
	105	104.905084	$\frac{5}{2}^+$	22.33%		111	110.9051107	$\frac{9}{2}^+$	2.8049d
	106	105.9034831	0^+	27.33%		112	111.9055333	1^+	14.97m
	107	106.9051285	$\frac{5}{2}^+$	6.5My		113	112.9040612	$\frac{9}{2}^+$	4.3%
	108	107.9038945	0^+	26.46%		114	113.9049168	1^+	71.9s
	109	108.9059535	$\frac{5}{2}^+$	13.7012h		115	114.9038783	$\frac{9}{2}^+$	95.7%
	110	109.9051524	0^+	11.72%		116	115.90526	1^+	14.10s
	111	110.907644	$\frac{5}{2}^+$	23.4m		117	116.9045157	$\frac{9}{2}^+$	43.2m
	112	111.9073133	0^+	21.03h		118	117.9063546	1^+	5.0s
	113	112.9101513	$(\frac{5}{2})^+$	93s		119	118.9058463	$\frac{9}{2}^+$	2.4m
	114	113.9103653	0^+	2.42m		120	119.9079615	1^+	3.08s
47 Ag	101	100.9128021	$\frac{9}{2}^+$	11.1m		121	120.9078488	$\frac{9}{2}^+$	23.1s
	102	101.912	5^+	12.9m		122	121.9102771	1^+	1.5s
	103	102.9089725	$\frac{7}{2}^+$	65.7m	50 Sn	107	106.9156667	$(\frac{5}{2}^+)$	2.90m
	104	103.9086282	5^+	69.2m		108	107.9119653	0^+	10.30m
	105	104.9065282	$\frac{1}{2}^-$	41.29d		109	108.9112869	$\frac{5}{2}^{(+)}$	18.0m
	106	105.9066664	1^+	23.96m		110	109.9078527	0^+	4.11h
	107	106.905093	$\frac{1}{2}^-$	51.839%		111	110.9077354	$\frac{7}{2}^+$	35.3m
	108	107.9059537	1^+	2.37m		112	111.9048208	0^+	0.97%
	109	108.9047555	$\frac{1}{2}^-$	48.161%		113	112.9051734	$\frac{1}{2}^+$	115.09d
	110	109.9061105	1^+	24.6s		114	113.9027818	0^+	0.65%
	111	110.9052947	$\frac{1}{2}^-$	7.45d		115	114.903346	$\frac{1}{2}^+$	0.36%
	112	111.9070041	$2^{(-)}$	3.130h		116	115.9017441	0^+	14.53%
	113	112.9065657	$\frac{1}{2}^-$	5.37h		117	116.9029538	$\frac{1}{2}^+$	7.68%
	114	113.9088079	1^+	4.6s		118	117.9016063	0^+	24.22%
	115	114.9087623	$\frac{1}{2}^-$	20.0m		119	118.9033089	$\frac{1}{2}^+$	8.58%
48 Cd	102	101.9147773	0^+	5.5m		120	119.9021966	0^+	32.59%
	103	102.913419	$(\frac{5}{2})^+$	7.3m		121	120.9042369	$\frac{3}{2}^+$	27.06h
	104	103.9098481	0^+	57.7m		122	121.9034401	0^+	4.63%
	105	104.9094678	$\frac{5}{2}^+$	55.5m		123	122.9057219	$\frac{11}{2}^-$	129.2d
	106	105.906458	0^+	1.25%		124	123.9052746	0^+	5.79%
	107	106.9066142	$\frac{5}{2}^+$	6.50h		125	124.9077849	$\frac{11}{2}^-$	9.64d
	108	107.9041834	0^+	0.89%		126	125.907654	0^+	0.1My
	109	108.9049856	$\frac{5}{2}^+$	462.6d		127	126.910351	$(\frac{11}{2}^-)$	2.10h
	110	109.9030056	0^+	12.49%		128	127.910535	0^+	59.1m
	111	110.9041816	$\frac{1}{2}^+$	12.8%		129	128.91344	$(\frac{3}{2}^+)$	2.16m
	112	111.9027572	0^+	24.13%	51 Sb	115	114.9065988	$\frac{5}{2}^+$	32.1m
	113	112.9044009	$\frac{1}{2}^+$	12.22%		116	115.9067972	3^+	15.8m
	114	113.9033581	0^+	28.73%		117	116.9048396	$\frac{5}{2}^+$	2.80h

(continued)

Z	A	Mass (u)	J^π	Abundance or Half-life	Z		A	Mass (u)	J^π	Abundance or Half-life
	118	117.9055319	1^+	3.6m	54	Xe	118	117.9165709	0^+	6m
	119	118.9039465	$\frac{5}{2}^+$	38.19h			119	118.9155543	$(\frac{5}{2}^+)$	5.8m
	120	119.9050743	1^+	15.89m			120	119.912152	0^+	40m
	121	120.903818	$\frac{5}{2}^+$	57.36%			121	120.9113865	$\frac{5}{2}^{(+)}$	40.1m
	122	121.9051754	2^-	2.7238d			122	121.9085484	0^+	20.1h
	123	122.9042157	$\frac{7}{2}^+$	42.64%			123	122.9084707	$(\frac{1}{2})^+$	2.08h
	124	123.9059375	3^-	60.20d			124	123.9058958	0^+	0.1%
	125	124.9052478	$\frac{7}{2}^+$	2.7582y			125	124.9063982	$(\frac{1}{2})^+$	16.9h
	126	125.9072482	$(8)^-$	12.46d			126	125.9042689	0^+	0.09%
	127	126.9069146	$\frac{7}{2}^+$	3.85d			127	126.9051796	$(\frac{1}{2}^-)$	36.4d
	128	127.9091673	8^-	9.01h			128	127.9035304	0^+	1.91%
	129	128.9091501	$\frac{7}{2}^+$	4.40h			129	128.9047795	$\frac{1}{2}^+$	26.4%
52	Te 114	113.912057	0^+	15.2m			130	129.9035079	0^+	4.1%
	115	114.9115786	$\frac{7}{2}^+$	5.8m			131	130.9050819	$\frac{3}{2}^+$	21.2%
	116	115.9084203	0^+	2.49h			132	131.9041545	0^+	26.9%
	117	116.9086342	$\frac{1}{2}^+$	62m			133	132.9059057	$\frac{3}{2}^+$	5.243d
	118	117.9058252	0^+	6.00d			134	133.9053945	0^+	10.4%
	119	118.9064081	$\frac{1}{2}^+$	16.03h			135	134.9072075	$\frac{3}{2}^+$	9.14h
	120	119.9040199	0^+	0.095%			136	135.9072195	0^+	8.9%
	121	120.9049298	$\frac{1}{2}^+$	16.78d			137	136.9115629	$\frac{7}{2}^-$	3.818m
	122	121.9030471	0^+	2.59%			138	137.9139885	0^+	14.08m
	123	122.904273	$\frac{1}{2}^+$	0.905%	55	Cs	126	125.909448	1^+	1.64m
	124	123.9028195	0^+	4.79%			127	126.9074176	$\frac{1}{2}^+$	6.25h
	125	124.9044247	$\frac{1}{2}^+$	7.12%			128	127.9077479	1^+	3.62m
	126	125.9033055	0^+	18.93%			129	128.9060634	$\frac{1}{2}^+$	32.06h
	127	126.9052173	$\frac{3}{2}^+$	9.35h			130	129.9067062	1^+	29.21m
	128	127.9044614	0^+	31.7%			131	130.9054602	$\frac{5}{2}^+$	9.689d
	129	128.9065956	$\frac{3}{2}^+$	69.6m			132	131.9064298	2^+	6.479d
	130	129.9062228	0^+	33.87%			133	132.9054469	$\frac{7}{2}^+$	100%
	131	130.9085219	$\frac{3}{2}^+$	25.0m			134	133.9067134	4^+	2.0648y
	132	131.9085238	0^+	3.204d			135	134.9059719	$\frac{7}{2}^+$	2.3My
	133	132.9109391	$(\frac{3}{2}^+)$	12.5m			136	135.9073057	5^+	13.16d
	134	133.9115405	0^+	41.8m			137	136.9070835	$\frac{7}{2}^+$	30.07y
	135	134.9164508	$(\frac{7}{2}^-)$	19.0s			138	137.9110105	3^-	33.41m
53	I 119	118.9101808	$\frac{5}{2}^+$	19.1m			139	138.9133579	$\frac{7}{2}^+$	9.27m
	120	119.9100478	2^-	81.0m			140	139.9172771	1^-	63.7s
	121	120.9073661	$\frac{5}{2}^+$	2.12h			141	140.920044	$\frac{7}{2}^+$	24.94s
	122	121.9075925	1^+	3.63m			142	141.9242923	0^-	1.70s
	123	122.9055979	$\frac{5}{2}^+$	13.27h			143	142.9273303	$\frac{3}{2}^+$	1.78s
	124	123.9062114	2^-	4.18d			144	143.9320274	1	1.01s
	125	124.9046242	$\frac{5}{2}^+$	59.408d			145	144.9353882	$\frac{3}{2}^+$	0.594s
	126	125.9056194	2^-	13.11d	56	Ba	125	124.9146202	$\frac{1}{2}^{(+)}$	3.5m
	127	126.9044684	$\frac{5}{2}^+$	100%			126	125.9112441	0^+	100m
	128	127.9058053	1^+	24.99m			127	126.9111213	$(\frac{1}{2}^+)$	12.7m
	129	128.9049875	$\frac{7}{2}^+$	15.7My			128	127.9083089	0^+	2.43d
	130	129.906674	5^+	12.36h			129	128.9086737	$\frac{1}{2}^+$	2.23h
	131	130.9061242	$\frac{7}{2}^+$	8.02070d			130	129.9063105	0^+	0.106%
	132	131.9079945	4^+	2.295h			131	130.9069308	$\frac{1}{2}^+$	11.50d
	133	132.9078065	$\frac{7}{2}^+$	20.8h			132	131.9050562	0^+	0.101%
	134	133.9098766	$(4)^+$	52.5m			133	132.9060024	$\frac{1}{2}^+$	10.52y
	135	134.9100503	$\frac{7}{2}^+$	6.57h						

Z	A	Mass (u)	J^π	Abundance or Half-life	Z	A	Mass (u)	J^π	Abundance or Half-life
	134	133.9045033	0^+	2.42%		145	144.9145069	$\frac{7}{2}^+$	5.984h
	135	134.9056827	$\frac{3}{2}^+$	6.593%		146	145.917588	$(2)^-$	24.15m
	136	135.9045701	0^+	7.85%		147	146.918979	$(\frac{3}{2}^+)$	13.4m
	137	136.9058214	$\frac{3}{2}^+$	11.23%		148	147.9221832	1^-	2.27m
	138	137.9052413	0^+	71.7%	60 Nd	136	135.9150205	0^+	50.65m
	139	138.9088354	$\frac{7}{2}^-$	83.06m		137	136.9146397	$\frac{1}{2}^+$	38.5m
	140	139.9105995	0^+	12.752d		138	137.91193	0^+	5.04h
	141	140.9144064	$\frac{3}{2}^-$	18.27m		139	138.9119242	$\frac{3}{2}^+$	29.7m
	142	141.9164482	0^+	10.6m		140	139.9093098	0^+	3.37d
	143	142.9206172	$\frac{5}{2}^-$	14.33s		141	140.9096048	$\frac{3}{2}^+$	2.49h
	144	143.9229405	0^+	11.5s		142	141.9077186	0^+	27.13%
57 La	131	130.9101085	$\frac{3}{2}^+$	59m		143	142.9098096	$\frac{7}{2}^-$	12.18%
	132	131.9101104	2^-	4.8h		144	143.9100826	0^+	23.8%
	133	132.9083964	$\frac{5}{2}^+$	3.912h		145	144.9125688	$\frac{7}{2}^-$	8.3%
	134	133.9084896	1^+	6.45m		146	145.9131121	0^+	17.19%
	135	134.906971	$\frac{5}{2}^+$	19.5h		147	146.9160958	$\frac{5}{2}^-$	10.98d
	136	135.9076512	1^+	9.87m		148	147.9168885	0^+	5.76%
	137	136.9064657	$\frac{7}{2}^+$	0.06My		149	148.9201442	$\frac{5}{2}^-$	1.728h
	138	137.9071068	5^+	0.0902%		150	149.9208866	0^+	5.64%
	139	138.9063482	$\frac{7}{2}^+$	99.9098%		151	150.9238247	$(\frac{3}{2})^+$	12.44m
	140	139.9094726	3^-	1.6781d		152	151.9246824	0^+	11.4m
	141	140.910957	$(\frac{7}{2}^+)$	3.92h	61 Pm	137	136.920713	$\frac{11}{2}^-$	2.4m
	142	141.9140745	2^-	91.1m		138	137.919445	1^+	10s
	143	142.9160586	$(\frac{7}{2})^+$	14.2m		139	138.9167598	$(\frac{5}{2})^+$	4.15m
	144	143.9195917	(3^-)	40.8s		140	139.9158016	1^+	9.2s
58 Ce	132	131.91149	0^+	3.51h		141	140.9136066	$\frac{5}{2}^+$	20.90m
	133	132.91155	$\frac{9}{2}^-$	4.9h		142	141.9129507	1^+	40.5s
	134	133.9090264	0^+	3.16d		143	142.9109276	$\frac{5}{2}^+$	265d
	135	134.9091456	$\frac{1}{2}(^+)$	17.7h		144	143.9125858	5^-	363d
	136	135.9071436	0^+	0.19%		145	144.9127439	$\frac{5}{2}^+$	17.7y
	137	136.9077776	$\frac{3}{2}^+$	9.0h		146	145.9146922	3^-	5.53y
	138	137.9059856	0^+	0.25%		147	146.9151339	$\frac{7}{2}^+$	2.6234y
	139	138.9066466	$\frac{3}{2}^+$	137.640d		148	147.9174678	1^-	5.370d
	140	139.905434	0^+	88.43%		149	148.9183292	$\frac{7}{2}^+$	53.08h
	141	140.9082711	$\frac{7}{2}^-$	32.501d		150	149.9209795	(1^-)	2.68h
	142	141.9092397	0^+	11.13%		151	150.9212027	$\frac{5}{2}^+$	28.40h
	143	142.9123812	$\frac{3}{2}^-$	33.039h		152	151.9234906	1^+	4.1m
	144	143.9136427	0^+	284.893d		153	152.9241132	$\frac{5}{2}^-$	5.4m
	145	144.9172279	$(\frac{3}{2})^-$	3.01m	62 Sm	138	137.92354	0^+	3.1m
	146	145.9186897	0^+	13.52m		139	138.922302	$(\frac{1}{2})^+$	2.57m
	147	146.922511	$(\frac{5}{2}^-)$	56.4s		140	139.918991	0^+	14.82m
59 Pr	136	135.9126469	2^+	13.1m		141	140.9184685	$\frac{1}{2}^+$	10.2m
	137	136.9106784	$\frac{5}{2}^+$	1.28h		142	141.9151933	0^+	72.49m
	138	137.9107489	1^+	1.45m		143	142.9146236	$\frac{3}{2}^+$	8.83m
	139	138.9089322	$\frac{5}{2}^+$	4.41h		144	143.9119947	0^+	3.1%
	140	139.9090712	1^+	3.39m		145	144.9134056	$\frac{7}{2}^-$	340d
	141	140.9076477	$\frac{5}{2}^+$	100%		146	145.9130368	0^+	103My
	142	141.9100399	2^-	19.12h		147	146.9148933	$\frac{7}{2}^-$	15%
	143	142.9108122	$\frac{7}{2}^+$	13.57d		148	147.9148179	0^+	11.3%
	144	143.9133006	0^-	17.28m		149	148.9171795	$\frac{7}{2}^-$	13.8%

(continued)

Z	A	Mass (u)	J^π	Abundance or Half-life	Z	A	Mass (u)	J^π	Abundance or Half-life
	150	149.9172715	0^+	7.4%		159	158.9253431	$\frac{3}{2}^+$	100%
	151	150.9199284	$\frac{5}{2}^-$	90y		160	159.927164	3^-	72.3d
	152	151.9197282	0^+	26.7%		161	160.9275663	$\frac{3}{2}^+$	6.88d
	153	152.9220939	$\frac{3}{2}^+$	46.27h		162	161.9294848	1^-	7.60m
	154	153.9222053	0^+	22.7%		163	162.9306439	$\frac{3}{2}^+$	19.5m
	155	154.9246359	$\frac{3}{2}^-$	22.3m	66 Dy	150	149.9255797	0^+	7.17m
	156	155.9255262	0^+	9.4h		151	150.9261796	$\frac{7}{2}^{(-)}$	17.9m
63 Eu	145	144.9162613	$\frac{5}{2}^+$	5.93d		152	151.9247139	0^+	2.38h
	146	145.9171997	4^-	4.59d		153	152.9257609	$\frac{7}{2}^{(-)}$	6.4h
	147	146.9167412	$\frac{5}{2}^+$	24.1d		154	153.924422	0^+	3.0My
	148	147.9181538	5^-	54.5d		155	154.925749	$\frac{3}{2}^-$	9.9h
	149	148.9179259	$\frac{5}{2}^+$	93.1d		156	155.9242783	0^+	0.06%
	150	149.9196983	$5^{(-)}$	35.8y		157	156.9254613	$\frac{3}{2}^-$	8.14h
	151	150.919846	$\frac{5}{2}^+$	47.8%		158	157.9244046	0^+	0.1%
	152	151.9217404	3^-	13.542y		159	158.9257357	$\frac{3}{2}^-$	144.4d
	153	152.9212262	$\frac{5}{2}^+$	52.2%		160	159.9251937	0^+	2.34%
	154	153.9229754	3^-	8.593y		161	160.9269296	$\frac{5}{2}^+$	18.9%
	155	154.9228894	$\frac{5}{2}^+$	4.7611y		162	161.9267947	0^+	25.5%
	156	155.9247509	0^+	15.19d		163	162.9287275	$\frac{5}{2}^-$	24.9%
	157	156.9254194	$\frac{5}{2}^+$	15.18h		164	163.9291712	0^+	28.2%
	158	157.9278419	(1^-)	45.9m		165	164.9316998	$\frac{7}{2}^+$	2.334h
	159	158.9290845	$\frac{5}{2}^+$	18.1m		166	165.9328032	0^+	81.6h
64 Gd	145	144.9216875	$\frac{1}{2}^+$	23.0m		167	166.935649	$(\frac{1}{2}^-)$	6.20m
	146	145.9183053	0^+	48.27d		168	167.93723	0^+	8.7m
	147	146.9190894	$\frac{7}{2}^-$	38.06h	67 Ho	157	156.9281881	$\frac{7}{2}^-$	12.6m
	148	147.9181098	0^+	74.6y		158	157.9289457	5^+	11.3m
	149	148.9193364	$\frac{7}{2}^-$	9.28d		159	158.9277085	$\frac{7}{2}^-$	33.05m
	150	149.9186555	0^+	1.79My		160	159.9287257	5^+	25.6m
	151	150.9203443	$\frac{7}{2}^-$	124d		161	160.9278517	$\frac{7}{2}^-$	2.48h
	152	151.9197879	0^+	0.2%		162	161.9290924	1^+	15.0m
	153	152.9217463	$\frac{3}{2}^-$	241.6d		163	162.9287303	$\frac{7}{2}^-$	4570y
	154	153.9208623	0^+	2.18%		164	163.9302306	1^+	29m
	155	154.9226188	$\frac{3}{2}^-$	14.8%		165	164.9303192	$\frac{7}{2}^-$	100%
	156	155.9221196	0^+	20.47%		166	165.9322813	0^-	26.83h
	157	156.9239567	$\frac{3}{2}^-$	15.65%		167	166.9331262	$\frac{7}{2}^-$	3.1h
	158	157.9241005	0^+	24.84%		168	167.9354964	3^+	2.99m
	159	158.9263851	$\frac{3}{2}^-$	18.479h		169	168.9368683	$\frac{7}{2}^-$	4.7m
	160	159.9270506	0^+	21.86%	68 Er	154	153.9327773	0^+	3.73m
	161	160.9296657	$\frac{5}{2}^-$	3.66m		155	154.9332043	$\frac{7}{2}^-$	5.3m
	162	161.9309812	0^+	8.4m		156	155.931015	0^+	19.5m
	163	162.93399	$(\frac{5}{2}^-)$	68s		157	156.9319455	$\frac{3}{2}^-$	18.65m
65 Tb	149	148.9232416	$\frac{1}{2}^+$	4.118h		158	157.929912	0^+	2.24h
	150	149.9236542	(2^-)	3.48h		159	158.9306807	$\frac{3}{2}^-$	36m
	151	150.9230982	$\frac{1}{2}^{(+)}$	17.609h		160	159.9290789	0^+	28.58h
	152	151.9240713	2^-	17.5h		161	160.9300013	$\frac{3}{2}^-$	3.21h
	153	152.9234309	$\frac{5}{2}^+$	2.34d		162	161.9287749	0^+	0.14%
	154	153.9246862	0	21.5h		163	162.9300293	$\frac{5}{2}^-$	75.0m
	155	154.9235004	$\frac{3}{2}^+$	5.32d		164	163.929197	0^+	1.61%
	156	155.9247437	3^-	5.35d		165	164.9307228	$\frac{5}{2}^-$	10.36h
	157	156.9240212	$\frac{3}{2}^+$	99y		166	165.93029	0^+	33.6%
	158	157.9254103	3^-	180y		167	166.9320454	$\frac{7}{2}^+$	22.95%

Z	A	Mass (u)	J^π	Abundance or Half-life	Z	A	Mass (u)	J^π	Abundance or Half-life
	168	167.9323678	0^+	26.8%		177	176.943755	$\frac{7}{2}^+$	6.734d
	169	168.9345881	$\frac{1}{2}^-$	9.40d		178	177.9459514	$1(^+)$	28.4m
	170	169.9354603	0^+	14.9%		179	178.9473242	$\frac{7}{2}(^+)$	4.59h
	171	170.9380259	$\frac{5}{2}^-$	7.516h		180	179.94988	$(3)^+$	5.7m
	172	171.9393521	0^+	49.3h		181	180.95197	$(\frac{7}{2}^+)$	3.5m
	173	172.9424	$(\frac{7}{2}^-)$	1.4m	72 Hf	167	166.9426	$(\frac{5}{2}^-)$	2.05m
69 Tm	161	160.933398	$\frac{7}{2}^+$	33m		168	167.94063	0^+	25.95m
	162	161.9339701	1^-	21.70m		169	168.9411586	$(\frac{5}{2})^-$	3.24m
	163	162.9326476	$\frac{1}{2}^+$	1.810h		170	169.93965	0^+	16.01h
	164	163.933451	1^+	2.0m		171	170.94049	$(\frac{7}{2}^+)$	12.1h
	165	164.9324325	$\frac{1}{2}^+$	30.06h		172	171.939458	0^+	1.87y
	166	165.9335531	2^+	7.70h		173	172.94065	$\frac{1}{2}^-$	23.6h
	167	166.9328488	$\frac{1}{2}^+$	9.25d		174	173.9400402	0^+	0.162%
	168	167.9341704	3^+	93.1d		175	174.941503	$\frac{5}{2}^-$	70d
	169	168.9342111	$\frac{1}{2}^+$	100%		176	175.9414018	0^+	5.206%
	170	169.9357979	1^-	128.6d		177	176.94322	$\frac{7}{2}^-$	18.606%
	171	170.9364258	$\frac{1}{2}^+$	1.92y		178	177.9436977	0^+	27.297%
	172	171.9383961	2^-	63.6h		179	178.9458151	$\frac{9}{2}^+$	13.629%
	173	172.9396003	$(\frac{1}{2}^+)$	8.24h		180	179.9465488	0^+	35.1%
	174	173.9421646	$(4)^-$	5.4m		181	180.9490991	$\frac{1}{2}^-$	42.39d
70 Yb	160	159.93756	0^+	4.8m		182	181.9505529	0^+	9My
	161	160.937853	$\frac{3}{2}^-$	4.2m	73 Ta	175	174.94365	$\frac{7}{2}^+$	10.5h
	162	161.93575	0^+	18.87m		176	175.9447406	$(1)^-$	8.09h
	163	162.9362655	$\frac{3}{2}^-$	11.05m		177	176.9444718	$\frac{7}{2}^+$	56.56h
	164	163.93452	0^+	75.8m		178	177.9457503	1^+	9.31m
	165	164.9353976	$\frac{5}{2}^-$	9.9m		179	178.9459341	$\frac{7}{2}^+$	1.82y
	166	165.9338796	0^+	56.7h		180	179.9474657	1^+	0.012%
	167	166.9349469	$\frac{5}{2}^-$	17.5m		181	180.9479963	$\frac{7}{2}^+$	99.988%
	168	167.9338945	0^+	0.13%		182	181.9501524	3^-	114.43d
	169	168.9351871	$\frac{7}{2}^+$	32.026d		183	182.9513732	$\frac{7}{2}^+$	5.1d
	170	169.9347587	0^+	3.05%		184	183.9540093	(5^-)	8.7h
	171	170.9363223	$\frac{1}{2}^-$	14.3%	74 W	174	173.94616	0^+	31m
	172	171.9363777	0^+	21.9%		175	174.94677	$(\frac{1}{2}^-)$	35.2m
	173	172.9382068	$\frac{5}{2}^-$	16.12%		176	175.94559	0^+	2.5h
	174	173.9388581	0^+	31.8%		177	176.94662	$(\frac{1}{2}^-)$	135m
	175	174.9412725	$\frac{7}{2}^-$	4.185d		178	177.9458484	0^+	21.6d
	176	175.9425684	0^+	12.7%		179	178.9470717	$(\frac{7}{2})^-$	37.05m
	177	176.9452571	$(\frac{9}{2}^+)$	1.911h		180	179.9467057	0^+	0.12%
	178	177.9466434	0^+	74m		181	180.9481981	$\frac{9}{2}^+$	121.2d
	179	178.95017	$(\frac{1}{2}^-)$	8.0m		182	181.9482055	0^+	26.3%
71 Lu	167	166.9383071	$\frac{7}{2}^+$	51.5m		183	182.9502245	$\frac{1}{2}^-$	14.28%
	168	167.9386986	(6^-)	5.5m		184	183.9509326	0^+	30.7%
	169	168.9376488	$\frac{7}{2}^+$	34.06h		185	184.9534206	$\frac{3}{2}^-$	75.1d
	170	169.9384722	0^+	2.00d		186	185.9543622	0^+	28.6%
	171	170.9379099	$\frac{7}{2}^+$	8.24d		187	186.9571584	$\frac{3}{2}^-$	23.72h
	172	171.9390822	4^-	6.70d		188	187.958487	0^+	69.4d
	173	172.9389269	$\frac{7}{2}^+$	1.37y		189	188.9619122	$(\frac{3}{2}^-)$	11.5m
	174	173.9403335	$(1)^-$	3.31y	75 Re	179	178.949981	$(\frac{5}{2})^+$	19.5m
	175	174.9407679	$\frac{7}{2}^+$	97.41%		180	179.9507877	$(1)^-$	2.44m
	176	175.9426824	7^-	2.59%					

(continued)

Z	A	Mass (u)	J^π	Abundance or Half-life	Z	A	Mass (u)	J^π	Abundance or Half-life
	181	180.9500646	$\frac{5}{2}^+$	19.9h		190	189.9599301	0^+	0.01%
	182	181.9512114	7^+	64.0h		191	190.9616847	$\frac{3}{2}^-$	2.9d
	183	182.9508213	$\frac{5}{2}^+$	70.0h		192	191.9610352	0^+	0.79%
	184	183.9525243	$3(^-)$	38.0d		193	192.9629845	$\frac{1}{2}^-$	50y
	185	184.9529557	$\frac{5}{2}^+$	37.4%		194	193.9626636	0^+	32.9%
	186	185.9549865	1^-	90.64h		195	194.9647744	$\frac{1}{2}^-$	33.8%
	187	186.9557508	$\frac{5}{2}^+$	62.6%		196	195.9649349	0^+	25.3%
	188	187.9581123	1^-	16.98h		197	196.9673234	$\frac{1}{2}^-$	18.3h
	189	188.9592284	$\frac{5}{2}^+$	24.3h		198	197.967876	0^+	7.2%
	190	189.9618161	$(2)^-$	3.1m		199	198.9705762	$\frac{5}{2}^-$	30.80m
76 Os	178	177.9533482	0^+	5.0m	79 Au	191	190.9636492	$\frac{3}{2}^+$	3.18h
	179	178.953951	$(\frac{1}{2}^-)$	6.5m		192	191.9648101	1^-	4.94h
	180	179.952351	0^+	21.5m		193	192.9641317	$\frac{3}{2}^+$	17.65h
	181	180.9532745	$\frac{1}{2}^-$	105m		194	193.9653389	1^-	38.02h
	182	181.9521862	0^+	22.10h		195	194.9650179	$\frac{3}{2}^+$	186.098d
	183	182.95311	$\frac{9}{2}^+$	13.0h		196	195.9665513	2^-	6.183d
	184	183.9524908	0^+	0.02%		197	196.9665516	$\frac{3}{2}^+$	100%
	185	184.954043	$\frac{1}{2}^-$	93.6d		198	197.9682252	2^-	2.6935d
	186	185.9538384	0^+	1.58%		199	198.968748	$\frac{3}{2}^+$	3.139d
	187	186.9557479	$\frac{1}{2}^-$	1.6%		200	199.9707179	$1(^-)$	48.4m
	188	187.955836	0^+	13.3%		201	200.9716408	$\frac{3}{2}^+$	26m
	189	188.9581449	$\frac{3}{2}^-$	16.1%		202	201.9737884	(1^-)	28.8s
	190	189.9584452	0^+	26.4%		203	202.9751373	$\frac{3}{2}^+$	53s
	191	190.960928	$\frac{9}{2}^-$	15.4d	80 Hg	188	187.967555	0^+	3.25m
	192	191.961479	0^+	41%		189	188.968132	$\frac{3}{2}^-$	7.6m
	193	192.9641481	$\frac{3}{2}^-$	30.5h		190	189.966277	0^+	20.0m
	194	193.9651793	0^+	6.0y		191	190.9670631	$(\frac{3}{2}^-)$	49m
77 Ir	183	182.956814	$\frac{5}{2}^-$	58m		192	191.965572	0^+	4.85h
	184	183.9573883	5^-	3.09h		193	192.9666442	$\frac{3}{2}^-$	3.80h
	185	184.95659	$\frac{5}{2}^-$	14.4h		194	193.9653818	0^+	520y
	186	185.9579511	5^+	16.64h		195	194.966639	$\frac{1}{2}^-$	9.9h
	187	186.9573608	$\frac{3}{2}^+$	10.5h		196	195.9658148	0^+	0.15%
	188	187.958852	1^-	41.5h		197	196.9671953	$\frac{1}{2}^-$	64.14h
	189	188.9587165	$\frac{3}{2}^+$	13.2d		198	197.9667518	0^+	9.97%
	190	189.9605923	$(4)^+$	11.78d		199	198.9682625	$\frac{1}{2}^-$	16.87%
	191	190.9605912	$\frac{3}{2}^+$	37.3%		200	199.9683087	0^+	23.1%
	192	191.9626022	$4(^+)$	73.831d		201	200.9702853	$\frac{3}{2}^-$	13.1%
	193	192.9629237	$\frac{3}{2}^+$	62.7%		202	201.9706256	0^+	29.86%
	194	193.9650756	1^-	19.15h		203	202.9728571	$\frac{5}{2}^-$	46.612d
	195	194.9659768	$\frac{3}{2}^+$	2.5h		204	203.9734756	0^+	6.87%
	196	195.9683799	(0^-)	52s		205	204.9760561	$\frac{1}{2}^-$	5.2m
	197	196.9696365	$\frac{3}{2}^+$	5.8m		206	205.9774987	0^+	8.15m
78 Pt	181	180.963177	$\frac{1}{2}^-$	51s	81 Tl	195	194.96965	$\frac{1}{2}^+$	1.16h
	182	181.9612676	0^+	2.2m		196	195.970515	2^-	1.84h
	183	182.961729	$\frac{1}{2}^-$	6.5m		197	196.9695362	$\frac{1}{2}^+$	2.84h
	184	183.959895	0^+	17.3m		198	197.9704663	2^-	5.3h
	185	184.9607538	$(\frac{9}{2}^+)$	70.9m		199	198.9698138	$\frac{1}{2}^+$	7.42h
	186	185.9594323	0^+	2.0h		200	199.9709454	2^-	26.1h
	187	186.960558	$\frac{3}{2}^-$	2.35h		201	200.9708038	$\frac{1}{2}^+$	72.912h
	188	187.9593957	0^+	10.2d		202	201.9720906	2^-	12.23d
	189	188.9608319	$\frac{3}{2}^-$	10.87h		203	202.9723291	$\frac{1}{2}^+$	29.524%

Z	A	Mass (u)	J^π	Abundance or Half-life
	204	203.9738486	2^-	3.78y
	205	204.9744123	$\frac{1}{2}^+$	70.476%
	206	205.9760953	0^-	4.199m
	207	206.9774079	$\frac{1}{2}^+$	4.77m
82 Pb	195	194.974471	$\frac{3}{2}^-$	15m
	196	195.97271	0^+	37m
	197	196.97338	$\frac{3}{2}^-$	8m
	198	197.97198	0^+	2.40h
	199	198.9729094	$\frac{3}{2}^-$	90m
	200	199.9718156	0^+	21.5h
	201	200.9728466	$\frac{5}{2}^-$	9.33h
	202	201.9721438	0^+	0.0525My
	203	202.9733755	$\frac{5}{2}^-$	51.873h
	204	203.9730288	0^+	1.4%
	205	204.9744671	$\frac{5}{2}^-$	15.3My
	206	205.974449	0^+	24.1%
	207	206.9758806	$\frac{1}{2}^-$	22.1%
	208	207.9766359	0^+	52.4%
	209	208.9810748	$\frac{9}{2}^+$	3.253h
	210	209.9841731	0^+	22.3y
	211	210.9887315	$\frac{9}{2}^+$	36.1m
	212	211.9918875	0^+	10.64h
	213	212.9965	$(\frac{9}{2}^+)$	10.2m
83 Bi	200	199.978142	7^+	36.4m
	201	200.9769707	$\frac{9}{2}^-$	108m
	202	201.9776745	5^+	1.72h
	203	202.9768681	$\frac{9}{2}^-$	11.76h
	204	203.9778052	6^+	11.22h
	205	204.9773747	$\frac{9}{2}^-$	15.31d
	206	205.9784829	$6(^+)$	6.243d
	207	206.9784552	$\frac{9}{2}^-$	31.55y
	208	207.9797267	$(5)^+$	0.368My
	209	208.9803832	$\frac{9}{2}^-$	100%
	210	209.9841049	1^-	5.013d
	211	210.9872581	$\frac{9}{2}^-$	2.14m
	212	211.9912715	$1(^-)$	60.55m
	213	212.9943748	$\frac{9}{2}^-$	45.59m
	214	213.9986987	1^-	19.9m
84 Po	201	200.982209	$\frac{3}{2}^-$	15.3m
	202	201.980704	0^+	44.7m
	203	202.9814129	$\frac{5}{2}^-$	36.7m
	204	203.9803071	0^+	3.53h
	205	204.9811654	$\frac{5}{2}^-$	1.66h
	206	205.9804652	0^+	8.8d
	207	206.9815782	$\frac{5}{2}^-$	5.80h
	208	207.9812311	0^+	2.898y
	209	208.9824158	$\frac{1}{2}^-$	102y
	210	209.9828574	0^+	138.376d
	211	210.9866369	$\frac{9}{2}^+$	0.516s
	212	211.9888518	0^+	0.299μs
	213	212.9928425	$\frac{9}{2}^+$	4.2μs
	214	213.9951859	0^+	164.3μs
	215	214.9994146	$\frac{9}{2}^+$	1.781ms
85 At	205	204.9860364	$\frac{9}{2}^-$	26.2m
	206	205.9865992	$(5)^+$	30.0m
	207	206.9857759	$\frac{9}{2}^-$	1.80h
	208	207.9865825	6^+	1.63h
	209	208.9861587	$\frac{9}{2}^-$	5.41h
	210	209.9871313	$(5)^+$	8.1h
	211	210.9874808	$\frac{9}{2}^-$	7.214h
	212	211.9907347	(1^-)	0.314s
	213	212.9929212	$\frac{9}{2}^-$	125ns
	214	213.9963564	1^-	558ns
	215	214.9986412	$\frac{9}{2}^-$	0.10ms
	216	216.0024088	$1(^-)$	0.30ms
	217	217.0047096	$\frac{9}{2}^-$	32.3ms
86 Rn	204	203.991365	0^+	1.24m
	205	204.991668	$\frac{5}{2}^-$	2.8m
	206	205.99016	0^+	5.67m
	207	206.9907268	$\frac{5}{2}^-$	9.25m
	208	207.9896312	0^+	24.35m
	209	208.9903766	$\frac{5}{2}^-$	28.5m
	210	209.9896799	0^+	2.4h
	211	210.9905854	$\frac{1}{2}^-$	14.6h
	212	211.9906889	0^+	23.9m
	213	212.9938684	$(\frac{9}{2}^+)$	25.0ms
	214	213.9953463	0^+	0.27μs
	215	214.9987292	$\frac{9}{2}^+$	2.30μs
	217	217.0039146	$\frac{9}{2}^+$	0.54ms
	218	218.0055863	0^+	35ms
	219	219.0094748	$\frac{5}{2}^+$	3.96s
	221	221.015455	$\frac{7}{2}(^+)$	25m
	222	222.0175705	0^+	3.8235d
	223	223.02179	$\frac{7}{2}$	23.2m
	224	224.02409	0^+	107m
	225	225.02844	$\frac{7}{2}$	4.5m
	226	226.03089	0^+	6.0m
87 Fr	207	206.9968594	$\frac{9}{2}^-$	14.8s
	208	207.9971338	7^+	59.1s
	209	208.9959154	$\frac{9}{2}^-$	50.0s
	210	209.9963983	6^+	3.18m
	211	210.9955293	$\frac{9}{2}^-$	3.10m
	212	211.996195	5^+	20.0m
	213	212.9961748	$\frac{9}{2}^-$	34.6s
	214	213.9989547	(1^-)	5.0ms
	215	215.000326	$\frac{9}{2}^-$	86ns
	217	217.0046165	$\frac{9}{2}^-$	22μs
	218	218.0075633	(1^-)	1.0ms

(continued)

Z	A	Mass (u)	J^π	Abundance or Half-life	Z	A	Mass (u)	J^π	Abundance or Half-life
	219	219.0092408	$\frac{9}{2}^-$	20ms		229	229.0334961	$(\frac{3}{2}^+)$	58m
	221	221.0142457	$\frac{5}{2}^-$	4.9m		230	230.0339274	0^+	20.8d
	222	222.017544	2^-	14.2m		231	231.0362892	$(\frac{5}{2}^-)$	4.2d
	223	223.0197307	$\frac{3}{2}(^-)$	21.8m		232	232.0371463	0^+	68.9y
	224	224.0232355	$1(^-)$	3.30m		233	233.0396282	$\frac{5}{2}^+$	0.1592My
	225	225.0256069	$\frac{3}{2}^-$	4.0m		234	234.0409456	0^+	0.0055%
88 Ra	218	218.0071239	0^+	25.6µs		235	235.0439231	$\frac{7}{2}^-$	0.72%
	219	219.0100688	$(\frac{7}{2})^+$	10ms		236	236.0455619	0^+	23.42My
	220	220.0110147	0^+	25ms		237	237.048724	$\frac{1}{2}^+$	6.75d
	221	221.0139078	$\frac{5}{2}^+$	28s		238	238.0507826	0^+	99.2745%
	222	222.0153618	0^+	38.0s		239	239.0542878	$\frac{5}{2}^+$	23.45m
	223	223.0184971	$\frac{3}{2}^+$	11.435d		240	240.0565857	0^+	14.1h
	224	224.020202	0^+	3.66d	93 Np	233	233.0407324	$(\frac{5}{2}^+)$	36.2m
	225	225.0236045	$\frac{1}{2}^+$	14.9d		234	234.0428886	(0^+)	4.4d
	226	226.0254026	0^+	1600y		235	235.0440559	$\frac{5}{2}^+$	396.1d
	227	227.0291707	$\frac{3}{2}^+$	42.2m		236	236.0465597	(6^-)	0.154My
	228	228.0310641	0^+	5.75y		237	237.0481673	$\frac{5}{2}^+$	2.14My
	229	229.0348203	$\frac{5}{2}(^+)$	4.0m		238	238.0509405	2^+	2.117d
	230	230.0370848	0^+	93m		239	239.0529314	$\frac{5}{2}^+$	2.3565d
89 Ac	223	223.019126	$(\frac{5}{2}^-)$	2.10m		240	240.0561688	(5^+)	61.9m
	225	225.0232206	$(\frac{3}{2}^-)$	10.0d		241	241.0582463	$(\frac{5}{2}^+)$	13.9m
	226	226.0260898	1^-	29h		242	242.061635	6^-	5.5m
	227	227.027747	$\frac{3}{2}^-$	21.773y	94 Pu	234	234.0433047	0^+	8.8h
	229	229.0329309	$(\frac{3}{2}^+)$	62.7m		235	235.0452815	$(\frac{5}{2}^+)$	25.3m
	230	230.0360251	(1^+)	122s		236	236.0460481	0^+	2.858y
	231	231.0385515	$(\frac{1}{2}^+)$	7.5m		237	237.0484038	$\frac{7}{2}^-$	45.2d
90 Th	223	223.0207952	$(\frac{5}{2})^+$	0.60s		238	238.0495534	0^+	87.7y
	224	224.0214593	0^+	1.05s		239	239.0521565	$\frac{1}{2}^+$	24110y
	225	225.0239414	$(\frac{3}{2})^+$	8.72m		240	240.0538075	0^+	6564y
	226	226.0248907	0^+	30.9m		241	241.0568453	$\frac{5}{2}^+$	14.35y
	227	227.0276989	$(\frac{1}{2}^+)$	18.72d		242	242.0587368	0^+	0.3733My
	228	228.0287313	0^+	1.9131y		243	243.061997	$\frac{7}{2}^+$	4.956h
	229	229.0317553	$\frac{5}{2}^+$	7340y		244	244.0641977	0^+	80.8My
	230	230.0331266	0^+	0.07538My		245	245.0677387	$(\frac{9}{2}^-)$	10.5h
	231	231.0362971	$\frac{5}{2}^+$	25.52h		246	246.0701984	0^+	10.84d
	232	232.0380504	0^+	100%	95 Am	237	237.0499707	$\frac{5}{2}(^-)$	73.0m
	233	233.0415769	$\frac{1}{2}^+$	22.3m		238	238.0519778	1^+	98m
	234	234.0435955	0^+	24.10d		239	239.0530185	$(\frac{5}{2})^-$	11.9h
	235	235.0475044	$(\frac{1}{2}^+)$	7.1m		240	240.0552878	(3^-)	50.8h
91 Pa	227	227.0287932	$(\frac{5}{2}^-)$	38.3m		241	241.0568229	$\frac{5}{2}^-$	432.2y
	228	228.0310369	(3^+)	22h		242	242.059543	1^-	16.02h
	229	229.0320886	$(\frac{5}{2}^+)$	1.50d		243	243.0613727	$\frac{5}{2}^-$	7370y
	230	230.0345326	(2^-)	17.4d		244	244.0642794	(6^-)	10.1h
	231	231.0358789	$\frac{3}{2}^-$	32760y		245	245.0664454	$(\frac{5}{2})^+$	2.05h
	232	232.0385817	(2^-)	1.31d		246	246.0697684	(7^-)	39m
	233	233.0402402	$\frac{3}{2}^-$	26.967d		247	247.072086	$(\frac{5}{2})$	23.0m
	234	234.0433023	4^+	6.70h					
	235	235.0454368	$(\frac{3}{2}^-)$	24.5m	96 Cm	243	243.0613822	$\frac{5}{2}^+$	29.1y
92 U	226	226.0293398	0^+	0.5s		244	244.0627463	0^+	18.10y
	227	227.0311401	$(\frac{3}{2}^+)$	1.1m					

Z	A	Mass u	J^π	Abundance or Half-life	Z	A	Mass (u)	J^π	Abundance or Half-life
	245	245.0654856	$\frac{7}{2}^+$	8500y		254	254.088016	(7^+)	275.7d
	246	246.0672176	0^+	4730y		255	255.0902664	$(\frac{7}{2}^+)$	39.8d
	247	247.0703468	$\frac{9}{2}^-$	15.6My	100 Fm	251	251.0815665	$(\frac{9}{2}^-)$	5.30h
	248	248.0723422	0^+	0.340My		252	252.0824601	0^+	25.39h
	249	249.0759471	$\frac{1}{2}(^+)$	64.15m		253	253.0851763	$\frac{1}{2}^+$	3.00d
	250	250.0783507	0^+	9000y		254	254.0868478	0^+	3.240h
	251	251.0822779	$(\frac{1}{2}^+)$	16.8m		255	255.0899555	$\frac{7}{2}^+$	20.07h
97 Bk	243	243.0630016	$(\frac{3}{2}^-)$	4.5h		256	256.0917665	0^+	157.6m
	244	244.0651679	(1^-)	4.35h		257	257.0950986	$(\frac{9}{2}^+)$	100.5d
	245	245.0663554	$\frac{3}{2}^-$	4.94d		258	258.097069	0^+	370µs
	246	246.0686668	$2(^-)$	1.80d	101 Md	255	255.0910752	$(\frac{7}{2}^-)$	27m
	247	247.0702985	$(\frac{3}{2}^-)$	1380y		256	256.0940528	$(0^-, 1^-)$	76m
	248	248.07308	(6^+)	9y		257	257.0955346	$(\frac{7}{2}^-)$	5.3h
	249	249.0749799	$\frac{7}{2}^+$	320d		258	258.0984253	(8^-)	55d
	250	250.0783105	2^-	3.217h		259	259.100503	$(\frac{7}{2}^-)$	103m
98 Cf	246	246.0687988	0^+	35.7h	102 No	253	253.090649	$(\frac{9}{2}^-)$	1.7m
	247	247.070992	$(\frac{7}{2}^+)$	3.11h		254	254.0909487	0^+	55s
	248	248.0721781	0^+	333.5d		255	255.0932324	$(\frac{1}{2}^+)$	3.1m
	249	249.0748468	$\frac{9}{2}^-$	351y		256	256.0942759	0^+	3.3s
	250	250.0764	0^+	13.08y		257	257.0968528	$(\frac{7}{2}^+)$	25s
	251	251.0795801	$\frac{1}{2}^+$	898y		258	258.0982	0^+	1.2ms
	252	252.0816196	0^+	2.645y		259	259.101024	$(\frac{9}{2}^+)$	58m
	253	253.0851268	$(\frac{7}{2}^+)$	17.81d	103 Lr	252	252.09533		1s
	254	254.0873162	0^+	60.5d		253	253.095258		1.3s
99 Es	249	249.076405	$\frac{7}{2}(^+)$	102.2m		257	257.099606	$(\frac{9}{2}^+)$	0.646s
	250	250.078654	(6^+)	8.6h		258	258.101883		4.3s
	251	251.0799836	$(\frac{3}{2}^-)$	33h		259	259.10299		5.4s
	252	252.0829722	(5^-)	471.7d		260	260.105320		180s
	253	253.084818	$\frac{7}{2}^+$	20.47d					

Bibliography

J. N. Bahcall, *Neutrino Astrophysics*. Cambridge: Cambridge University Press, 1989.

M. G. Bowler, *Nuclear Physics*. Oxford: Pergamon Press, 1974.

W. E. Burcham, *Elements of Nuclear Physics*. London: Longman, 1979.

W. E. Burcham, *Nuclear Physics: An Introduction,* 2nd edition. London: Longman, 1973.

B. L. Cohen, *Concepts of Nuclear Physics*. New York: McGraw-Hill Book Company, 1971.

W. N. Cottingham and D. A. Greenwood, *An Introduction to Nuclear Physics*. Cambridge: Cambridge University Press, 1986.

A. Das and T. Ferbel, *Introduction to Nuclear and Particle Physics*. New York: Wiley, 1994.

L. R. B. Elton, *Introductory Nuclear Theory: Second Edition*. Philadelphia: W.B. Saunders Company, 1966.

H. A. Enge, *Introduction to Nuclear Physics*. Reading, MA: Addison-Wesley Publishing Company, Inc., 1966.

H. Frauenfelder and E. M. Henley, *Sub-atomic Physics*. Englewood Cliffs: Prentice-Hall, 1974.

K. L. G. Heyde, *Basic Ideas and Concepts in Nuclear Physics: An Introductory Approach*. Bristol: Institute of Physics Publishing, 1994.

N. A. Jelley, *Fundamentals of Nuclear Physics*. Cambridge: Cambridge University Press, 1990.

K. S. Krane, *Introductory Nuclear Physics*. New York: Wiley, 1988.

W. E. Meyerhof, *Elements of Nuclear Physics*. New York: McGraw-Hill, 1967.

E. B. Paul, *Nuclear and Particle Physics*. Amsterdam: North-Holland Publishing Company, 1969.

M. A. Preston, *Physics of the Nucleus*. Reading, MA: Addison-Wesley Publishing Company, 1962.

E. Segrè, *Nuclei and Particles,* 2nd edition. Reading, MA: Benjamin Cummings Publishing Company, 1977.

W. S. C. Williams, *Nuclear and Particle Physics*. Oxford: Clarendon Press, 1991.

Index

Ablation, 189–190
Allowed transitions, 116–120
Alpha decay, 94–104
  decay rate for, 103
  Geiger-Nutall rule for, 96–97
  lifetime of, 97–101
  Q value for, 94–96
  selection rules for, 102–104
Alpha particle, 19, 94, 142
  angular momentum of, 102–103
  binding energy of, 95
  parity of, 102
  scattering of, 19
Angular momentum, 10, 211
  barrier, 102–104
  conservation of, 9–10, 102, 211
  orbital, 43, 56, 62, 71–73, 102, 211
  spin, 43–44, 56, 72–73, 211
  total, 56, 62, 71–73, 153
Antineutrino, 8, 106, 201–202, 254
Antisymmetric wave function, 63, 218
Associated Legendre polynomials, 51

Atomic
  energy levels, 42–45
  mass, 4
  mass unit (u), 4
  number, 4
  weight, 5
Auger
  electron, 248
  process, 128, 248

Barn (unit), 5
Baryon, 7, 199, 211–212, 231
  number, 10, 203
    conservation, 10, 204, 215
Beta decay, 32–37, 106–127, 205, 222
  allowed, 116–120
  double, 122–127, 253
  energy spectrum of, 108–109, 113
  Fermi theory of, 109–114
  forbidden, 115
  $f\tau$-value for, 118–120
  parity conservation in, 117

Beta decay (*continued*)
  parity violation in, 120–122
  Q value for, 107
  superallowed, 120
Binding energy
  electronic, 26–27, 107, 136
  nuclear, 26–31, 45–46, 95
B mesons, 215
Boson, 7, 199
  gauge, 7, 199, 202, 228, 232–233.
    See also $W^\pm$ boson, Z^0 boson
Bottom, 10
  quark, 209
  mesons, 215
Branching
  fraction, 86–87
  ratio, 86, 103, 136, 231
Breakeven, 192
Breeder reactor, 172–173
Breit-Wigner formula, 153, 156
Bremsstrahlung, 185–186

Cabibbo
  angle, 232
  theory, 231–232, 253
CANDU reactor, 172
Carbon-Nitrogen-Oxygen (CNO) cycle,
    182–183
Center of mass system, 147
Chain reaction, 164, 167–168
Charm, 10, 209
  quark, 209
Charmonium, 214
Charge
  conjugation, 210, 241
  conservation, 10
  distribution, 19–23
Cherenkov radiation, 246, 250
Collective model, 75–80
Color, 218–221, 238–239
Configurational mixing, 75
Confinement, 215
  inertial, 189–194
  magnetic, 187–189
  time, 186, 191
Conservation laws, 9–10, 232–233,
    245–246

Control rod, 169–170
Coulomb
  barrier, 98, 155, 160, 175, 178
  effects, 155–156
  factor, 114
  interaction, 13, 98, 230
  scattering, 16
  term, 29–30, 161
CPT theorem, 243
CP violation, 241–243
Critical
  mass, 167–168
  radius, 168
Cross section
  differential scattering, 18
  electron-positron, 239
  neutron, 47–48, 150–155, 165–167,
    170
  scattering, 17
  uranium, 166–167

Daughter, 83
Decay
  alpha, 94–104
  beta, 32–37, 106–127, 205
  constant, 83–84, 86–91
  electromagnetic, 234–235
  gamma, 128–139
  hadronic, 234–235
  hadronic lepton, 234
  leptonic, 234
  leptonic hadron, 234
  multimodal, 86–87
  nonleptonic hadron, 234–235
  rate, 83–84, 86–91
  semileptonic hadron, 234–235
  sequential, 87–88
Degeneracy, 43–44, 53
Delayed neutrons, 164, 171
Density of states, 110–111, 152–153
Deuteron, 142, 146, 176
  stripping reaction, 148–150
Differential scattering cross section,
    18
Dirac neutrino, 253
Direct reaction, 152
Doppler broadening, 155, 171

Double beta decay, 122–127, 253
   neutrinoless, 253
   two neutrino, 122
Down quark, 209
d-t fusion, 176, 179–181

Eccentricity, 161
Eightfold way, 217
Einstein relation, 204
Electric
   dipole, 129–130
      moment, 67–68
   monopole moment, 67–68
   multipole, 131–134
   octupole moment, 67–68
   quadrupole, 131
      moment, 67–71
Electromagnetic
   interaction, 9, 43, 202, 225,
      232–233
   moments, 67–75
Electron, 8, 200
   capture, 33, 106–107
   wave function, 110
Electron-positron annihilation, 227,
   239
Electroweak theory, 228, 245
Endothermic, 95, 148
Endpoint energy, 108, 256–257
Excited states, 47, 49–50, 143
   of hadrons, 211
   parity of, 66–67
   of proton, 208
   rotational, 76–78
   in the shell model, 65–67, 76–77
   spin of, 66–67
   vibrational, 78–80
Exothermic, 95, 148

Fermi
   decay, 116
   energy, 58
   factor, 114
   golden rule, 109
   plot, 114–116
   theory of beta decay, 109–114
Fermi-Dirac statistics, 7

Fermi-Kurie plot, 114–116
Fermilab, 214
Fermion, 7, 43
Fermi (unit), 5
Feynman diagram, 202–206, 232–238
Fissile, 163
Fission, 158–173
   barrier, 160
   induced, 162
   reactor, 171–173
   spontaneous, 160
   yield, 164
Flavor, 209, 253–256
Forbidden transitions, 115–120
Form factor, 20–21
Fragmentation, 240
Frequency doubler, 190
Fourier transform, 89
$f\tau$-values, 118–120
Fusion, 175–184
   inertial confinement, 189–193
   laser, 189
   magnetic confinement, 187–189
   reactor, 187–195
      inertial confinement, 189–193
      magnetic confinement, 187–189
      pinch machines, 187

GALEX, 249–250
Gamma decay, 128–139
   selection rules for, 135–136
Gamma ray, 128
Gauge boson, 7, 199, 202, 228,
   232–233
Gamow-Teller decay, 116
Geiger-Nutall rule, 96–97
Gell-Mann-Nishijima relation, 217
g-factor, 71–73
Gluon, 202–203, 216, 218, 232, 238,
   240
Golden rule. *See* Fermi, golden rule
Grand unified theories (GUTs),
   245–258
Graviton, 202
Gravity, 9, 202
Ground state, 62–65
   of hadrons, 211

Hadron, 7, 207, 209–215
Hadronic interaction. *See* Strong
     interaction
Halflife, 84
Hamiltonian, 57
Harmonic oscillator, 55, 61
Heisenberg uncertainty principle, 89, 151
Hole state, 67
Homestake, 248–249
Hydrogen-like atom, 42–43

Impact parameter, 16–19
Incident channel, 151
Induced fission, 162
Inertial confinement, 189–194
Infinite square well, 50–54
Internal conversion, 128, 136–139
     coefficient, 138–139
Internet sites, 6
Ionization energy, 44–45
Isobar, 5
Isospin, 10, 216–217
Isotone, 5, 47–48
Isotope, 4, 13, 47
Isotopic spin. *See* Isospin

J/ψ meson, 214

Kamiokande, 250
Kaon. *See* K meson
K meson, 241–243
Kurie plot. *See* Fermi-Kurie plot

Laboratory frame, 143, 147, 229
Landé *g*-factor, 71
Lawson
     criterion, 187
     parameter, 187, 191–193
Lepton, 7, 199–202, 215, 233–234
     generation, 10, 199, 201, 215, 245
     number, 10, 201, 203, 245–246
Lifetime, 84–86
Liquid drop model, 26–40, 45, 75

Magic numbers
     atomic, 44
     nuclear, 49, 53, 58

Magnetic
     confinement, 187–189
     dipole, 130
       moment, 68–69, 71–75
     monopole, 67–68
     multipole, 132, 134–135
     quadrupole moment, 67–68
Majorana neutrino, 253
Mass, 5–6, 10
     distribution, 23–25
     excess, 27
     neutrino, 107, 245, 253, 256–258
     number, 4
     parabola, 32
     quark, 209, 216
     reduced, 99, 152
Matrix element, 110, 118, 133
Maxwell-Boltzmann distribution,
     179–180
Mean free path, 168
Meson, 7, 207, 210–211, 213–215,
     233–226, 239–241
Meteorite, age of, 91–92
Mikheyev-Smirnov-Wolfenstein (MSW)
     effect, 253
Mirror nuclei, 67–68
Mixing angle, 232
Moderator, 168–170
Moment of inertia, 77–78
Mössbauer effect, 129
Multipole
     moment, 67, 131, 133
     radiation, 131–135
Muon, 200, 239, 254, 257–258
     neutrino, 200–202, 254, 257–258

Natural abundance, 5, 14
Neutrino, 8, 106, 199
     atmospheric, 254
     Dirac, 253
     electron, 8, 32, 106, 247–258
     Majorana, 253
     mass, 107, 245, 253, 256–258
     muon, 200–202, 254, 257–258
     oscillations, 252–256
     solar, 247–252
     tau, 200, 234, 258

Neutron, 4, 8, 142
  cross section, 47–48, 150–155,
      165–167, 170
  delayed, 164, 171
  epithermal, 154
  fast, 47, 154
  g-factor for, 73
  magnetic moment of, 73–74, 208
  prompt, 163, 167
  reactions, 150–155
  separation energy, 37–40, 149, 153
  slow, 154
  thermal, 154
Nonfissile, 163
Nuclear
  binding energy, 26–31, 45–46, 95
  mass, 4, 5
  radius, 47
  spin, 62–64
Nuclear magneton, 73
Nucleon, 4
  separation energy, 37–40, 149, 153
Nuclide, 4, 47

Off the mass shell, 204

Parent, 83
Paring term, 29–31
Parity, 10, 62–64, 211, 241–243
  conservation, 10, 102, 117
  non-conservation, 120–122
Partial
  decay rate, 87, 136
  width, 231
Pauli exclusion principle, 42–43
Phonon, 79
Photon, 135, 203, 232–233
  spin, 135
Pickup reaction, 146
Pi meson. *See* Pion
Pinch machines, 187
Pion, 201, 219
Positron, 8, 32
  annihilation, 227, 239
Prompt neutrons, 163, 167
Proton, 4, 8, 142
  decay, 246

g-factor, 73
  magnetic moment, 73–74
  separation energy, 37–40
Proton-proton
  collisions, 230
  cycle, 182
  fusion, 176

Quantum chromodynamics (QCD), 218
Quantum number
  good, 37, 58
  orbital angular momentum, 42–43
  principal, 42–43, 53
  for quarks, 210
  spin, 44, 53
Quark, 199, 207
  bottom, 209
  charm, 209
  color, 218
  down, 209
  flavor, 209
  generation, 215, 231
  mass, 209, 216
  number, 215, 245
  strange, 209
  top, 209, 214
  up, 209
Quark-gluon plasma, 216
Q value
  for alpha decay, 94–96
  for beta decay, 107
  for reactions, 147

Radiative capture, 165
Radioactive
  dating, 90–92
  waste, 163
Radioisotope, 4
Reaction
  compound nucleus, 142
  direct, 142
  endothermic, 147–148
  exothermic, 147–148
  pickup, 146
  Q value for, 147
  resonance, 142
  stripping, 146, 148–150

Reactor
    breeder, 172–173
    CANDU, 172
    fission, 171–173
    fusion, 184–195
Recoil energy, 96, 129
Reduced mass, 99, 152
Relativistic effects, 19, 114
Resonances, 144, 150, 153, 166–167,
    239
Rotational states, 76–78
Rutherford scattering, 16–19

SAGE, 249
Scattering, 141
    Rutherford, 16–17
    coulombic, 16
    elastic, 141, 144, 165, *See also*
        Rutherford scattering
    inelastic, 143–146, 165, 208
Schmidt lines, 73–75
Schrödinger equation, 3, 49, 51, 56, 79,
    88, 99, 104, 111
Second harmonic generator, 190
Self-conjugate, 210
Semiempirical mass formula, 27–31,
    158, 161–162
Separation energy, 37–40
    neutron, 37–40, 149
    proton, 37–40
Shape factor, 115–117
Shell model, 42–67, 75
Single particle model, 65–67
Solar neutrino unit (SNU), 249
Spectroscopic notation, 43, 53. 57, 63,
    210
Spherical Bessel functions, 52–53
Spherical harmonics, 51–52
Spin, 7, 62, 72–79, 116, 135, 199,
    201–202, 209–212, 216
Spin-orbit interaction, 56–58
Spontaneous fission, 160
Square well, 54–55
Strangeness, 10

Strange quark, 209
Stripping reaction, 146–150
Strong interaction, 9, 43, 202, 225, 233
Stellerator, 188
Sudbury neutrino observatory (SNO),
    255
Superallowed transitions, 120
Super Kamiokande , 250–251, 254
Surface term, 29–30, 161
Symmetric wave function, 63
Symmetry term, 29–30

Tau
    lepton, 200, 234, 258
    neutrino, 200, 234, 258
Threshold energy, 165–167
Time reversal, 243
Tokomak, 188
    spherical, 189
Top, 10
    quark, 209, 214
Tritium, 194, 257
Triton, 142, 177

Up quark, 209

Vibrational states, 78–80
Virtual particles, 204
Volume term, 28–30

$W^{\pm}$ boson, 202–205, 222, 225–231, 254
Weak
    boson, 202–205, 222, 225–231, 254
    interaction, 9, 202, 205, 222–225, 233
Web sites, 6
Weisskopf estimates, 133–135
Wentzel-Kramers-Brillouin (WKB)
    approximation, 100
Woods-Saxon distribution, 21–24, 54

X boson, 246

Z^0 boson, 202, 222, 227–231, 254
Zweig's rule, 237